Sascha Falahat

Investigation of the Reactions 25,26Mg (alpha,n) and 18O (alpha,n)

Sascha Falahat

Investigation of the Reactions 25,26Mg (alpha,n) and 18O (alpha,n)

Experimental investigation and impact on stellar nucleosynthesis

Südwestdeutscher Verlag für Hochschulschriften

Impressum/Imprint (nur für Deutschland/only for Germany)
Bibliografische Information der Deutschen Nationalbibliothek: Die Deutsche Nationalbibliothek verzeichnet diese Publikation in der Deutschen Nationalbibliografie; detaillierte bibliografische Daten sind im Internet über http://dnb.d-nb.de abrufbar.
Alle in diesem Buch genannten Marken und Produktnamen unterliegen warenzeichen-, marken- oder patentrechtlichem Schutz bzw. sind Warenzeichen oder eingetragene Warenzeichen der jeweiligen Inhaber. Die Wiedergabe von Marken, Produktnamen, Gebrauchsnamen, Handelsnamen, Warenbezeichnungen u.s.w. in diesem Werk berechtigt auch ohne besondere Kennzeichnung nicht zu der Annahme, dass solche Namen im Sinne der Warenzeichen- und Markenschutzgesetzgebung als frei zu betrachten wären und daher von jedermann benutzt werden dürften.

Verlag: Südwestdeutscher Verlag für Hochschulschriften GmbH & Co. KG
Dudweiler Landstr. 99, 66123 Saarbrücken, Deutschland
Telefon +49 681 37 20 271-1, Telefax +49 681 37 20 271-0
Email: info@svh-verlag.de

Approved by: Mainz, Universität, Diss., 2010

Herstellung in Deutschland:
Schaltungsdienst Lange o.H.G., Berlin
Books on Demand GmbH, Norderstedt
Reha GmbH, Saarbrücken
Amazon Distribution GmbH, Leipzig
ISBN: 978-3-8381-2466-7

Imprint (only for USA, GB)
Bibliographic information published by the Deutsche Nationalbibliothek: The Deutsche Nationalbibliothek lists this publication in the Deutsche Nationalbibliografie; detailed bibliographic data are available in the Internet at http://dnb.d-nb.de.
Any brand names and product names mentioned in this book are subject to trademark, brand or patent protection and are trademarks or registered trademarks of their respective holders. The use of brand names, product names, common names, trade names, product descriptions etc. even without a particular marking in this works is in no way to be construed to mean that such names may be regarded as unrestricted in respect of trademark and brand protection legislation and could thus be used by anyone.

Publisher: Südwestdeutscher Verlag für Hochschulschriften GmbH & Co. KG
Dudweiler Landstr. 99, 66123 Saarbrücken, Germany
Phone +49 681 37 20 271-1, Fax +49 681 37 20 271-0
Email: info@svh-verlag.de

Printed in the U.S.A.
Printed in the U.K. by (see last page)
ISBN: 978-3-8381-2466-7

Copyright © 2011 by the author and Südwestdeutscher Verlag für Hochschulschriften GmbH & Co. KG and licensors
All rights reserved. Saarbrücken 2011

1. Berichterstatter : Dr. Ulrich Ott
2. Berichterstatter : Prof. Dr. Klaus Wendt

Meinen Eltern und Meinem Bruder Saman

Zusammenfassung

In der vorliegenden Dissertation werden die Kernreaktionen ^{25}Mg(α,n)^{28}Si, ^{26}Mg(α,n)^{29}Si und ^{18}O(α,n)^{21}Ne im astrophysikalisch interessanten Energiebereich von $E_\alpha = 1000$ keV bis $E_\alpha = 2450$ keV untersucht.
Die Experimente wurden am Nuclear Structure Laboratory der University of Notre Dame (USA) mit dem vor Ort befindlichen Van-de-Graaff Beschleuniger KN durchgeführt. Hierbei wurden Festkörpertargets mit evaporiertem Magnesium oder anodisiertem Sauerstoff mit α-Teilchen beschossen und die freigesetzten Neutronen untersucht. Zum Nachweis der freigesetzten Neutronen wurde mit Hilfe von Computersimulationen ein Neutrondetektor basierend auf ^{3}He-Zählrohren konstruiert. Weiterhin wurden aufgrund des verstärkten Auftretens von Hintergrundreaktionen verschiedene Methoden zur Datenanalyse angewendet. Abschliessend wird mit Hilfe von Netzwerkrechnungen der Einfluss der Reaktionen ^{25}Mg(α,n)^{28}Si, ^{26}Mg(α,n)^{29}Si und ^{18}O(α,n)^{21}Ne auf die stellare Nukleosynthese untersucht.

Abstract

In the present dissertation, the nuclear reactions ^{25}Mg$(\alpha,$n$)^{28}$Si, ^{26}Mg$(\alpha,$n$)^{29}$Si and ^{18}O$(\alpha,$n$)^{21}$Ne are investigated in the astrophysically interesting energy region from $E_\alpha = 1000$ keV to $E_\alpha = 2450$ keV.

The experiments were performed at the Nuclear Structure Laboratory of the University of Notre Dame (USA) with the Van-de-Graaff accelerator KN. Solid state targets with evaporated magnesium or anodized oxygen were bombarded with α-particles and the released neutrons detected. For the detection of the released neutrons, computational simulations were used to construct a neutron detector based on ^{3}He counters. Because of the strong occurrence of background reactions, different methods of data analysis were employed. Finally, the impact of the reactions ^{25}Mg$(\alpha,$n$)^{28}$Si, ^{26}Mg$(\alpha,$n$)^{29}$Si and ^{18}O$(\alpha,$n$)^{21}$Ne on stellar nucleosynthesis is investigated by means of network calculations.

Contents

Zusammenfassung	i
Abstract	iii
Contents	v
List of Figures	viii
List of Tables	xi

1 Introduction 1

2 Nucleosynthesis in Stars 3
  2.1 The Classical S-Process 5
    2.1.1 S-Process Branchings 7
  2.2 Stellar Sites for the S-Process 9
    2.2.1 AGB Stars 9
    2.2.2 Massive Stars 10
  2.3 Nuclear Physics behind Nucleosynthesis 12
    2.3.1 The Astrophysical S-Factor 13
  2.4 Reaction Networks 15
  2.5 Observational Evidence - Meteoritic Grains 16

3 Previous Results 17
  3.1 ^{25}Mg(α,n)^{28}Si & ^{26}Mg(α,n)^{29}Si 18
  3.2 ^{18}O(α,n)^{21}Ne 27

4 Experimental Techniques and Procedures 31
  4.1 The KN Accelerator and Beam Transport System 32
    4.1.1 RBS Beamline 32
    4.1.2 0° Beamline 34
  4.2 Neutron Detection 35
    4.2.1 Moderation of Neutrons 36
    4.2.2 Design Principles 36
    4.2.3 Computational Simulations 37
    4.2.4 Parameter Studies 40
    4.2.5 Design of the Test Detector 43
  4.3 Neutron Detector Construction 46
    4.3.1 Electronics Setup of the Neutron Detectors 50

		4.3.2 Validation of Computational Calculations	50

- 4.4 Counting Station 52
- 4.5 (p,γ)-Measurement Setup 52
- 4.6 Target Production 54
  - 4.6.1 Thermal Evaporation 54
  - 4.6.2 Backing Material 55
  - 4.6.3 Evaporation Process 56
  - 4.6.4 Vanadium Targets 59

5 Experimental Results   61
- 5.1 Neutron Detector Performance 61
  - 5.1.1 Positioning 61
  - 5.1.2 Efficiency 61
  - 5.1.3 Ring Ratio 69
- 5.2 RBS Measurements 72
- 5.3 Target Thickness Measurements 74
- 5.4 Background Correction Techniques 75
  - 5.4.1 Background Targets 78
  - 5.4.2 Artifical Background Correction 84
- 5.5 ^{25}Mg(α,n)^{28}Si 90
- 5.6 ^{26}Mg(α,n)^{29}Si 95
- 5.7 ^{18}O(α,n)^{21}Ne 99
- 5.8 Results from Ring Ratio Calculations 104
- 5.9 Reaction Rates 109

6 Impact Studies   115
- 6.1 ^{25}Mg(α,n)^{28}Si & ^{26}Mg(α,n)^{29}Si 116
- 6.2 ^{18}O(α,n)^{21}Ne 129

7 Conclusions   131

A Beam Tuning Procedures   133
- A.1 Energy Change 133
- A.2 Tuning Process for α-Beam 133
- A.3 Switch to Beam Profile Monitor (BPM) in Target Room 134

B ^{3}He Counter   135

C Correction Formalism   136

D Analytic Expressions for the Reaction Rates   138

E Nucleosynthesis Plots   140
- E.1 Comparison to SiC X Data 140
  - Ti 141
  - Sr 142
  - Zr 143
  - Mo 144
  - Ba 145

E.2 Abundance Evolution 146
   $T_9 = 1.5$ 147
   $T_9 = 1.7$ 152

Acknowledgements   **157**

Bibliography   **158**

Curriculum Vitae   **165**

List of Figures

2.1 The $\langle\sigma\rangle_{(A)}N_{s(A)}$ curve. 7
2.2 S-process path in the mass region A=147-149 8
2.3 S-process abundance distribution with updated Nd cross sections. 10
2.4 Energy generation of the advanced burning stages of a massive star 11

3.1 Previous results for the cross section of ^{25}Mg(α,n)^{28}Si 20
3.2 Previous results for the cross section of ^{26}Mg(α,n)^{29}Si 21
3.3 Previous results for the astrophysical S-factor of ^{25}Mg(α,n)^{28}Si 23
3.4 Previous results for the astrophysical S-factor of ^{26}Mg(α,n)^{29}Si 23
3.5 Presolar grain measurements and calculations of ^{26}Mg(α,n)^{29}Si 26
3.6 ^{25}Mg(α,n)^{28}Si and ^{26}Mg(α,n)^{29}Si during convective carbon burning 26
3.7 ^{18}O(α,n)^{21}Ne during convective carbon burning 27
3.8 Three isotope plots of the Ne isotopes. 28
3.9 Previous results for the astrophysical S-factor of ^{18}O(α,n)^{21}Ne 28
3.10 Previous results for the cross section of ^{18}O(α,n)^{21}Ne 29

4.1 The ^{18}O(α,n)^{21}Ne reaction at $E_\alpha = 1866$ keV 33
4.2 Cross section for the reaction ^{3}He(n,p)^{3}H 35
4.3 Comparison of data provided by ENDF and GEANT4 38
4.4 MCNP5 results for different HDPE matrices 42
4.5 GEANT4 results for different HDPE matrices 42
4.6 Drawing of the test detector 44
4.7 Different views of the test detector 45
4.8 Drawing of neutron detector 46
4.9 Drawing of neutron detector with different ring orientation 47
4.10 View on inner structure of the neutron detector 48
4.11 Pictures of the neutron detector 49
4.12 Electronics scheme for the neutron detector 50
4.13 Picture of the (p,γ) measurement setup 53
4.14 Sketch of a dimple boat 57
4.15 Picture of the MgOTa mixture 57
4.16 Sketch of a pinhole boat 58
4.17 Sketch of a tube heater 58
4.18 Picture of a Mg target after evaporation 59
4.19 Picture of the evaporation setup 60

5.1 Recorded neutron yield versus detector position 62
5.2 Neutron detector efficiency in comparison to GEANT4 results 64

LIST OF FIGURES

5.3   Neutron detector efficiency in comparison to MCNP results 65
5.4   Neutron detector efficiency up to E_n = 10 MeV in a linear plot 65
5.5   Neutron detector efficiency up to E_n = 10 MeV in a log-log plot 66
5.6   (n,n) scattering data compared to experimental and computational results . 67
5.7   Efficiency functions of various detection systems 68
5.8   Neutron detector efficiency as a function of the ring ratio 69
5.9   Schematic drawing of a compound reaction 70
5.10 Ring ratio as a function of the initial neutron energy 70
5.11 Ring ratio as a function of initial neutron energy up to E_n = 3.5 MeV . . . 71
5.12 RBS spectrum of a contaminated ^{26}Mg target 73
5.13 RBS spectrum of a clean ^{25}Mg target 73
5.14 Cross section for relevant reactions including background reactions 76
5.15 Yield curve for ^{26}Mg(α,n)^{29}Si for an enriched ^{26}Mg target 77
5.16 Yield curves of ^{18}O(α,n)^{21}Ne for various ^{18}O targets 79
5.17 Yield curves for ^{18}O(α,n)^{21}Ne at E_α = 1000 - 1300 keV 80
5.18 Yield curve of ^{18}O(α,n)^{21}Ne and Ta background data 81
5.19 Yield data for ^{25}Mg(α,n)^{28}Si, scaled ^{24}Mg data and corrected ^{25}Mg data . . 83
5.20 Illustration of different ^{13}C impurities 85
5.21 Schematic drawing of a Mg target 86
5.22 Yield curve of ^{26}Mg(α,n)^{29}Si and cross section data for ^{11}B(α,n)^{14}N 88
5.23 Yield curve of ^{26}Mg(α,n)^{29}Si and artificial background 89
5.24 Experimental yield curve and corrected yield curve for ^{25}Mg(α,n)^{28}Si 92
5.25 Cross section obtained for ^{25}Mg(α,n)^{28}Si 93
5.26 S-factor obtained for ^{25}Mg(α,n)^{28}Si 94
5.27 Cross section obtained for ^{26}Mg(α,n)^{29}Si 97
5.28 S-factor obtained for ^{26}Mg(α,n)^{29}Si 98
5.29 Obtained cross section for ^{18}O(α,n)^{21}Necompared to NACRE data 101
5.30 Obtained cross section for ^{18}O(α,n)^{21}Necompared to adjusted NACRE data 102
5.31 S-factor for ^{18}O(α,n)^{21}Ne 103
5.32 Yield curve of the ring ratio for ^{25}Mg(α,n)^{28}Si 105
5.33 Yield curve of the ring ratio for ^{26}Mg(α,n)^{29}Si 105
5.34 Yield curve of the ring ratio for ^{18}O(α,n)^{21}Ne 106
5.35 Reaction rate obtained for ^{25}Mg(α,n)^{28}Si 110
5.36 Reaction rate obtained for ^{26}Mg(α,n)^{29}Si 111
5.37 Reaction rate obtained for ^{18}O(α,n)^{21}Ne 113
5.38 Ratio of reaction rates of ^{18}O(α,n)^{21}Ne and ^{18}O(α,γ)^{22}Ne 114

6.1   Neutron density as a function of time for T_9 = 1.5 & 1.7 118
6.2   Isotopic abundance variations by ^{25}Mg(α,n)^{28}Si at T_9 = 1.5 121
6.3   Isotopic abundance variations by ^{25}Mg(α,n)^{28}Si at T_9 = 1.7 (1.4 d) 122
6.4   Isotopic abundance variations by ^{26}Mg(α,n)^{29}Si at T_9 = 1.5 123
6.5   Isotopic abundance variations by ^{26}Mg(α,n)^{29}Si at T_9 = 1.7 (1.4 d) 124
6.6   Isotopic abundance variations by ^{25,26}Mg(α,n)^{28,29}Si at T_9 = 1.5 125
6.7   Isotopic abundance variations by ^{25,26}Mg(α,n)^{28,29}Si at T_9 = 1.7 (1.4 d) . . 126

B.1   Schematic illustration of a ^{3}He counter 135

E.1   Ti abundance variations (T_9=1.5) compared to SiC X data 141
E.2   Ti abundance variations (T_9=1.7) compared to SiC X data 141

- E.3 Sr abundance variations (T_9=1.5) compared to SiC X data 142
- E.4 Sr abundance variations (T_9=1.7 after 1.4 days) compared to SiC X data . 142
- E.5 Zr abundance variations (T_9=1.5) compared to SiC X data 143
- E.6 Sr abundance variations (T_9=1.7 after 1.4 days) compared to SiC X data . 143
- E.7 Mo abundance variations (T_9=1.5) compared to SiC X data 144
- E.8 Mo abundance variations (T_9=1.7 after 1.4 days) compared to SiC X data . 144
- E.9 Ba abundance variations (T_9=1.5) compared to SiC X data 145
- E.10 Ba abundance variations (T_9=1.7 after 1.4 days) compared to SiC X data . 145
- E.11 Si abundance evolution at T_9=1.5 for NACREx1 147
- E.12 Si abundance evolution at T_9=1.5 for NACREx0.1 148
- E.13 Si abundance evolution at T_9=1.5 for NACREx0.01 148
- E.14 Zr abundance evolution at T_9=1.5 for NACREx1 149
- E.15 Zr abundance evolution at T_9=1.5 for NACREx0.1 149
- E.16 Zr abundance evolution at T_9=1.5 for NACREx0.01 150
- E.17 Mo abundance evolution at T_9=1.5 for NACREx1 150
- E.18 Mo abundance evolution at T_9=1.5 for NACREx0.1 151
- E.19 Mo abundance evolution at T_9=1.5 for NACREx0.01 151
- E.20 Si abundance evolution at T_9=1.7 for NACREx1 152
- E.21 Si abundance evolution at T_9=1.7 for NACREx0.1 153
- E.22 Si abundance evolution at T_9=1.7 for NACREx0.01 153
- E.23 Zr abundance evolution at T_9=1.7 for NACREx1 154
- E.24 Zr abundance evolution at T_9=1.7 for NACREx0.1 154
- E.25 Zr abundance evolution at T_9=1.7 for NACREx0.01 155
- E.26 Mo abundance evolution at T_9=1.7 for NACREx1 155
- E.27 Mo abundance evolution at T_9=1.7 for NACREx0.1 156
- E.28 Mo abundance evolution at T_9=1.7 for NACREx0.01 156

List of Tables

2.1	Burning energies for ^{18}O(α,n)^{21}Ne, ^{25}Mg(α,n)^{28}Si and ^{26}Mg(α,n)^{29}Si	14
2.2	Types of presolar grains in primitive meteorites and IDPs	16
4.1	List of reactions used for the KN energy calibration	32
4.2	Detection efficiency for differently sized HDPE matrices	41
4.3	Detection efficiency for different numbers of ^{3}He tubes	43
5.1	Resonances used for target thickness measurements	74
5.2	Target thickness and Mg:O ratios for the production targets.	74
5.3	Resolved resonance states for ^{25}Mg(α,n)^{28}Si	90
5.4	Obtained resonance parameters for ^{25}Mg(α,n)^{28}Si	91
5.5	Resolved resonances for ^{26}Mg(α,n)^{29}Si	95
5.6	Resonance parameters for ^{26}Mg(α,n)^{29}Si	96
5.7	Resonance parameters obtained for ^{18}O(α,n)^{21}Ne compared to NDS	99
5.8	Resonance parameters for ^{18}O(α,n)^{21}Ne in comparison to Denker	100
5.9	Nuclear structure parameters for compound nucleus mechanism	104
5.10	Neutron energies for the n$_0$ group of ^{18}O(α,n)^{21}Ne	107
5.11	Obtained mixing ratios for ^{18}O(α,n)^{21}Ne resonances	108
5.12	Temperatures at which ^{18}O(α,n)^{21}Ne starts to dominate over ^{18}O(α,γ)^{22}Ne	112
6.1	List of decay chains affecting final abundances	117
6.2	List of time scales for Mo(γ,n) reactions	119
6.3	Comparison of δ^iSi obtained and from Hoppe et al.	128
6.4	Isotopic ratios of Neon obtained and from Heck et al.	129
D.1	Fit parameters for the reaction rate of ^{25}Mg(α,n)^{28}Si	138
D.2	Fit parameters for the reaction rate of ^{26}Mg(α,n)^{29}Si	139
D.3	Fit parameters for the reaction rate of ^{18}O(α,n)^{21}Ne from Denker	139
D.4	Fit parameters for the reaction rate of ^{18}O(α,n)^{21}Ne from the present work	139
E.1	Normalization isotopes for SiC X data	140

Chapter 1

Introduction

In September 1920, Sir Arthur Eddington addressed the internal constitution of stars and several aspects of current star models. Models based on gravitational contraction of a star as the energy source powering the light emission did not agree with astronomical observations of stars, such as the lifetime. Eddington, lacking the future knowledge of nuclear physics, suggested that the "stars are the crucibles in which the lighter atoms which abound in the nebulae are compounded into more complex elements"[1]. After Eddington's address, an era of discoveries in nuclear physics and astrophysics set the foundation for Burbidge, Burbidge, Fowler and Hoyle's theory on stellar nucleosynthesis[2]. They were the first to categorize the synthesis of the chemical elements and their isotopes by the processes which occur inside the star that depend on the temperature, mass and density of the star.

The synthesis of the elements lighter than iron was mainly assigned to processes involving charged particle reactions. Since the large coulomb barrier would hinder charged particle reactions, the synthesis of the elements heavier than iron required a second type of reactions, so called neutron capture reactions. Two main neutron capture processes were identified : the slow neutron capture process (s-process) and the rapid neutron capture process (r-process). The first process is characterized by lower neutron densities ($N_n \simeq 10^8$ cm^{-3}) while the r-process operates at $N_n \geq 10^{20}$ cm^{-3}.

Cameron was the first to recognize that specific nuclear reactions could serve as neutron sources for the neutron capture reactions[3]. By considering the energy generation in stars and the nuclear physics for specific reactions, he identified the ^{13}C(α,n)^{16}O reaction as a main neutron source. Cameron only took early burning phases, such as the carbon cycle into consideration, whereas Fowler, Burbidge and Burbidge built on Cameron's work and considered different phases of stellar evolution[4]. In their work from 1955 they point out that if unprocessed hydrogen mixes into the expanding helium core of a star, other reactions could serve as additional neutron sources. The reactions ^{17}O(α,n)^{20}Ne, ^{21}Ne(α,n)^{24}Mg, ^{22}Ne(α,n)^{25}Mg, ^{25}Mg(α,n)^{28}Si and ^{26}Mg(α,n)^{29}Si were identified as additional neutron sources inside the star, each at different stages of the stellar evolution.

In particular, ^{17}O(α,n)^{20}Ne has an important role in recycling the neutrons captured by ^{16}O(n,γ)O^{17} and ^{20}Ne(n,γ)^{21}Ne, respectively[5]. The reactions ^{25}Mg(α,n)^{28}Si and ^{26}Mg(α,n)^{29}Si were identified as neutron sources during the neon burning phase in massive stars[6]. The role of ^{18}O(α,n)^{21}Ne for the nucleosynthesis of heavier elements seems marginal. However, the competition between ^{18}O(α,n)^{21}Ne and ^{18}O(α,γ)^{22}Ne is important for the synthesis of the neon isotopes and may effect later burning phases[7, 8].

The reactions ^{18}O(α,n)^{21}Ne, ^{25}Mg(α,n)^{28}Si and ^{26}Mg(α,n)^{29}Si have been experimentally investigated over the course of time, but the results do not deliver a conclusive picture regarding their role in stellar nucleosynthesis. To achieve a more conclusive picture one has to carefully determine these reaction cross sections and the reaction rate at the energies relevant for stellar nucleosynthesis. The determination of the cross section and reaction rate is especially difficult due to experimental challenges.

Experimentally determined data sets drive stellar models which highlight the role of specific nuclear reactions. The fundamental link between these models and their validation is provided by the measurement of isotopic abundances in meteoritic grains and by astronomical observations.

Recent results from meteoritic grain measurements and stellar models indicate that an enhanced experimental data set of the reactions ^{18}O$(\alpha,n)^{21}$Ne, ^{25}Mg$(\alpha,n)^{28}$Si and ^{26}Mg$(\alpha,n)^{29}$Si would unveil a clearer picture on their role toward stellar nucleosynthesis. Varying the rates within the uncertainties produces noticeable variations in the isotopic abundance patterns calculated by stellar nucleosynthesis models.

In this thesis, experimental challenges will be discussed and new measurements will be presented on the reactions ^{18}O$(\alpha,n)^{21}$Ne, ^{25}Mg$(\alpha,n)^{28}$Si and ^{26}Mg$(\alpha,n)^{29}$Si. An investigation towards the impact of the new reaction rates in the stellar models and a comparison with recent meteoritic grain measurements will conclude this thesis.

Chapter 2

Nucleosynthesis in Stars

As a star evolves, it ignites and burns through different phases, which, at first, are dominated by charged-particle reactions. Those charged-particle reactions do not only contribute to the overall energy generation of the star, but also lead up to the production of the iron peak elements.

Stars with initial masses M $\leq$ 8 M$_\odot$ experience hydrogen burning (H-Burning) and helium burning (He-Burning)[9]. For M $\geq$ 8 M$_\odot$ the temperature is high enough to also ignite carbon burning (C-Burning). Finally, stars with M $\geq$ 10-12 M$_\odot$ may activate additional burning phases such as neon burning (Ne-Burning), oxygen burning (O-Burning) and silicon burning (Si-Burning) before exploding as a core-collapse supernova[10].

As mentioned in the introduction, the synthesis of nuclei higher in mass than ^{56}Fe is mainly related to the appearance of neutron capture reactions (s- and r-process). These reactions are not hindered by the coulomb barrier and are favored over charged particle reactions for nuclei of masses above ^{56}Fe.

A smaller fraction of the elements beyond iron are produced by the so called proton capture process (p-process), especially on the proton rich side of the valley of stability, and by the νp-process[11, 12]. Both processes do not play a role for the scope of this thesis.

The general equation for a neutron capture reaction is :

$$(Z, A) + n \Rightarrow (Z, A+1) + \gamma \qquad (2.1)$$

where Z is the atomic and A the mass number[13]. If the isotope (Z, A+1) is stable, the following (n,γ) reaction leads to the isotope (Z, A+2) etc. By considering a chain of (n,γ) reactions, certain produced isotopes are unstable toward β^--decay, which is defined as :

$$(Z, A+1) \Rightarrow (Z+1, A+1) + e^- + \bar{\nu}_e \qquad (2.2)$$

The chain of (n,γ) reactions is then interrupted by the production of unstable nuclei or an (n,γ) $\rightleftharpoons$ (γ,n) equilibrium. To reach isotopes with higher atomic number, the β^--decay in equation 2.2 has to occur faster than the (n,γ) reaction. On the other hand, to reach isotopes with higher mass numbers, the (n,γ) reaction has to occur on a shorter time scale than the β^--decay.

Regarding the lifetimes of the specific interactions, one can distinguish between three processes :

- $\tau_\beta \gg \tau_n$

    The neutron capture reaction occurs on a shorter time scale than the β^--decay. Therefore, isotopes with a higher mass number are produced.

- $\tau_n \sim \tau_\beta$

  Both processes occur at a similar rate. This situation is encountered at so-called branching points.

- $\tau_\beta \ll \tau_n$

  Here the β^--decay occurs on a shorter time scale than the neutron capture. Consequently isobars with higher atomic numbers are produced.

In general, the first case is typical for the classical r-process while the third case is characteristic for the s-process. The processes are distinguished by the required neutron density which the nuclei need to be exposed to.

The extreme conditions for the r-process are matched in explosive stellar environments, while the s-process is operating during hydrostatic burning phases of a star. Both processes follow their own path on the chart of nuclides. One can find isotopes that are for example produced only by the s-process and are called s-only isotopes, while there are also r-only and s-r-isotopes (of mixed origin).

The scope of this thesis does not allow a detailed review of the r-process, while the s-process is the major process sensitive to the neutron releasing reactions ^{25}Mg$(\alpha,n)^{28}$Si, ^{26}Mg$(\alpha,n)^{29}$Si and ^{18}O$(\alpha,n)^{21}$Ne.

2.1 The Classical S-Process

An analytical study of the s-process was first performed by Clayton and collaborators in the 1960s, without taking into account the astrophysical conditions[14].
The abundance variation with time of an isotope along the s-process path may be written as follows :

$$\frac{dN_A(t)}{dt} = N_n(t)N_{A-1}\langle\sigma v\rangle_{A-1} - N_n(t)N_A(t)\langle\sigma v\rangle_A - \lambda_\beta(t)N_A(t) \quad (2.3)$$

where $N_n(t)$ is the neutron density, $\lambda_\beta = (ln2)/t_{1/2}$, N_A is the abundance at the time t and $\langle\sigma v\rangle_A$ is the reaction rate. Additionally, one has to take into account the time-dependence of the process as well. Not only the production of the isotope A, but also its destruction by β-decay and neutron capture are implemented. As a consequence, equation 2.3 consists of three terms :

- $N_n(t)N_{A-1}\langle\sigma v\rangle_{A-1}$:

    This term describes the increase in abundance of the isotope by neutron capture of its neighbor A-1.

- $N_n(t)N_A(t)\langle\sigma v\rangle_A$:

    By capturing a neutron, the abundance of the isotope A is decreased.

- $\lambda_\beta(t)N_A(t)$:

    As mentioned before, the competition process to the neutron capture, the β-decay also effects the abundance of the isotope as described in the last term. The last term is only present if isotope A is unstable.

Several assumptions have been made to derive an approximation of equation 2.3 and to establish a link to observed isotopic abundances.
The first assumption is that the unstable nuclei (described in term 3) are sufficiently short-lived so that the s-process can continue. The s-process operates for masses higher than iron in an neutron energy region of 10 - 100 keV. This implicates that the cross section for the neutron capture reactions is proportional to the inverse of the thermal velocity (E $\sim 1/v$). The reaction rate $\langle\sigma v\rangle_A$, in a first approximation, is therefore constant over the s-process relevant temperature range and can be written as :

$$\langle\sigma\rangle = \frac{\langle\sigma v\rangle}{v_T} \quad (2.4)$$

where the reaction rate can be expressed through the Maxwell-Boltzmann distribution (see section 2.3). Equation 2.3 can then be rewritten as:

$$\frac{dN_A(t)}{dt} = N_n(t)N_{A-1}\langle\sigma\rangle_{A-1}v_T - N_n(t)N_A(t)\langle\sigma\rangle_A v_T \quad (2.5)$$

By introducing the neutron exposure τ,

$$\tau = \int v_T n_n(t)dt \rightarrow d\tau = v_T N_n(t)dt \quad (2.6)$$

equation 2.5 can be expressed as :

$$\frac{dN_s(A)}{d\tau} = \sigma(A-1)N_s(A-1) - \sigma(A)N_s(A) \qquad (2.7)$$

The change in abundance depends on the product σN_s which varies smoothly with mass number. Clayton introduced this approximation known as local approximation[13, 14].
Seeger et al. were not able to reproduce the s-isotope abundances with one single neutron exposure[15]. The product of cross section and abundance is given by Seeger et al. in analytical form by :

$$\langle\sigma\rangle_{(A)} N_{s(A)} = \frac{G \cdot N_{56}^{\odot}}{\tau_0} \prod_{i=56}^{A} (1 + \frac{1}{\tau_0 \langle\sigma\rangle_{(i)}})^{-1} \qquad (2.8)$$

where $N_{56}^{\odot}$ is the observed solar ^{56}Fe abundance and τ_0, as well as G, are fit parameters. Therefore, in equation 2.8, the remaining input parameter left is the neutron capture cross section.
Plotted versus the mass number A, the so called $\langle\sigma\rangle_{(A)} N_{s(A)}$ curve plotted with the empirical data from the s-only nuclei shows a distinct behaviour at the neutron magic nuclei. The smooth shape of the curve between the neutron magic numbers indicates a close to equilibrium situation in this range (see figure 2.1).
G and τ_0, the two fit parameters from equation 2.8, that determine the shape of the $\langle\sigma\rangle_{(A)} N_{s(A)}$ curve, are the first indicators for stellar sites, especially with respect to seed abundance and neutron exposure. Ward et al. showed that at least two neutron exposures are necessary[16].
Each component is responsible for a specific mass region of nuclei and the nucleosynthesis associated to that mass region. The *weak* component of the s-process is responsible for the nucleosynthesis of mass numbers $60 \leq A \leq 90$ while the nucleosynthesis above $A \geq 90$ is predominatly associated with the *main* component[17, 18].
Finally, the *strong* component is required to explain about half of the solar ^{208}Pb abundances[19]. Once the s-process reaches the region of α-instability, it is terminated by the reaction chain ^{209}Bi(n,γ)^{210}Bi(γ,α)^{206}Pb.

2.1 The Classical S-Process

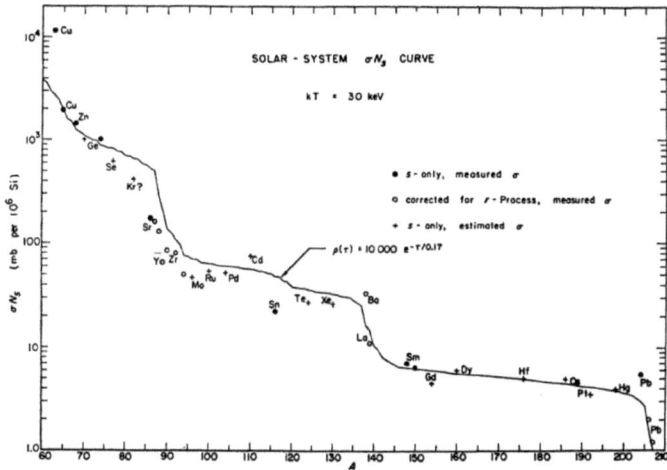

Figure 2.1: The $\langle\sigma\rangle_{(A)}N_{s(A)}$ curve plotted versus the mass number A. The solid line is calculated corresponding to the work of Seeger et al.[15].

2.1.1 S-Process Branchings

The competition between neutron captures and β^--decays is critical for determining the path of the s-process. Therefore, the condition $\tau_\beta \ll \tau_n$ represents the main characteristic for the path of the s-process.

Nevertheless, some nuclei exhibit comparable neutron capture and β-decay rates ($\tau_n \sim \tau_\beta$), possibly resulting in a branching of the reaction flow. These nuclei are described as possible branching points of the s-process.

The occurence of a branching can be observed by comparing the abundance of nucleus (Z, A+1) with the abundance of nucleus (Z+1, A). Differences in the observed abundances can lead to more detailed information about the physical environment, respectively the astrophysical site. For instance, s-process branching points can help determine the neutron density to which the seed nuclei had to be exposed to.

One can define the strength of a branching in terms of the β-decay rates λ_β of the involved nuclei :

$$f_\beta = \frac{\lambda_\beta}{\lambda_\beta + \lambda_n} \tag{2.9}$$

One can then calculate the neutron density n_n analytically via:

$$n_n = \frac{1-f_\beta}{f_\beta} \cdot \frac{1}{v_T \langle\sigma\rangle_i} \cdot \frac{ln(2)}{t^*_{1/2}(i)} \tag{2.10}$$

where i denotes the isotope at which the branching occurs.

As λ_β is dependent on the temperature, each branching point can be used to derive important information about the stellar environment. For different branching points,

different sets of parameters can be found, which constraint the astrophysical scenarios, that have to be taken into account[20].
The canonical approach has been successful in describing not only the s-process branching, but also the s-abundance distribution. However, as more (n,γ) cross sections were experimentally determined, the limitations of the canonical approach became evident.
The isotope ^{142}Nd is an s-process isotope and located at a drop of the $\langle\sigma\rangle_{(A)}N_{s(A)}$ curve. This drop is caused by the appearance of the neutron magic number N= 82 and the subsequent low neutron capture cross section.

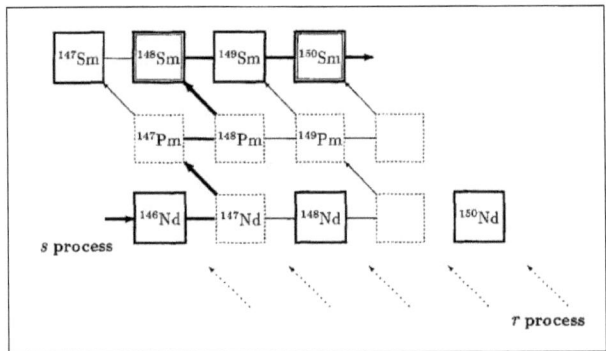

Figure 2.2: Illustrated is the s-process path in the mass region A=147-149. The isotopes ^{148}Sm and ^{150}Sm are shielded from the r-process and determine the branching strength[20]. A significant branching at A = 147 - 149 results in a higher $\langle\sigma\rangle_{(A)}N_{s(A)}$ value for ^{150}Sm compared to ^{148}Sm.

Wisshak et al. used their experimentally determined (n,γ) cross sections of the Nd isotopes to calculate the abundances of the Nd isotopes with the classical approach. Their results showed a clear overproduction of the isotope ^{142}Nd. The mismatch between the classical approach and observed abundances led to the conclusion that the classical approach is not accurate enough to describe the stellar scenario in which the s-process takes place[21]. Only nucleosynthesis calculations in realistic stellar models are able to reproduce in detail the production of the s-process elements.

2.2 Stellar Sites for the S-Process

The neutrons for the (n,γ)-reactions are supplied by (α,n) reactions on specific nuclei[3, 4]. These (α,n) reactions are known as stellar neutron sources. The primary stellar neutron sources are the reactions ^{13}C(α,n)^{16}O and ^{22}Ne(α,n)^{25}Mg. These reactions can take place during burning phases of the star in which helium is available.

Each component of the s-process needs to be assigned to different types of stars or their respective phases. This is mainly due to the different physical conditions needed to reproduce the abundance patterns observed.

Low mass asymptotic giant branch (AGB) stars (M $\leq$ 8 M$_\odot$) were determined to be the sites for the main and strong component[18, 22]. The weak component of the s-process is assigned to massive stars (M > 10 M$_\odot$)[23].

2.2.1 AGB Stars

A star reaches the AGB phase after helium burning has been exhausted in the core. At this point the He-Burning continues in a shell, together with a more external hydrogen burning shell.

During most of the AGB phase, the energy to sustain the stellar structure is supplied by the hydrogen shell. The temperature in the helium shell is not high enough to activate e.g. the triple-α process.

However, as the hydrogen shell processes more material, the temperature increases above the helium shell until helium burning is activated in a flash (Thermal Pulse, TP). This is mainly due to the high temperature dependence of the triple-α process.

The large energy generation during the TP induces a convective instability in the region between the helium shell and the hydrogen shell (He-intershell region). The He-intershell region expands by cooling and eventually reduces the hydrogen burning efficiency. The TP typically lasts few hundred years until the triple-α process loses efficiency, again.

The He-intershell contracts and becomes radiative again, while the H-shell returns to hydrostatic burning[9].

After the TP, the hydrogen shell may eventually be deactivated and the convective envelope may dredge up part of the He-intershell material. This causes mixing of fresh hydrogen below the previous hydrogen shell location (Third Dredge Up, TDU)[24].

The freshly supplied protons may be captured by ^{12}C producing ^{13}C through the ^{12}C(p,γ)^{13}N(e$^+$ ν)^{13}C reaction chain. This process is located in a small radiative region just below the hydrogen shell known as ^{13}C-pocket.

Within the ^{13}C-pocket, the reaction ^{13}C(α,n)^{16}O burns at T$_9$ $\sim$ 0.1 during the intershell phase, forming the s-process elements. The s-process enriched pocket is mixed into the He-intershell during the next TP. Following the TDU, the s-process elements may be dredged up into the envelope. The neutron exposure produced within the ^{13}C-pocket accounts for about 95 % of the produced s-processe nuclides whereas the partial activation of the reaction ^{22}Ne(α,n)^{25}Mg during the TP accounts for the remaining 5%[22].

In terms of processed material and its isotopic composition, it is important to note that the metallicity of the star and the profile of the ^{13}C-pocket remain as the crucial parameters. As Arlandini et al. were able to illustrate (see figure 2.3), AGB star models with solar-like metallicity provide the best agreement with the main component of the s-process[18].

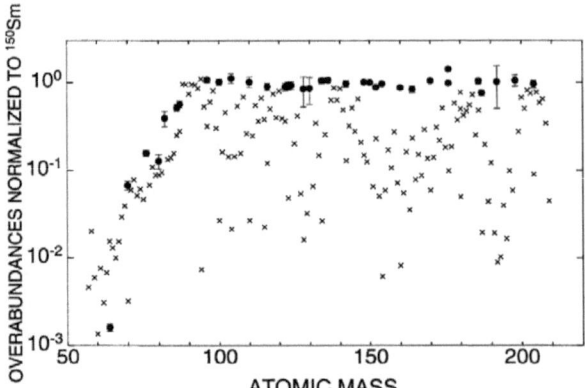

Figure 2.3: S-process abundance distribution that reproduces the solar system main s-component, with updated Nd cross sections. Obtained from Arlandini et al.[18].

2.2.2 Massive Stars

Massive stars, M $\geq$ 10 M$_\odot$, are responsible for the production of the weak s-process component[25].
During hydrogen burning ^{14}N is being synthesized from the initial CNO isotopes and represents approximately 2 % of the core composition.
^{14}N will be converted into ^{22}Ne via the reaction chain ^{14}N$(\alpha,\gamma)^{18}$F$(\beta^+)^{18}$O$(\alpha,\gamma)^{22}$Ne. At the end of the helium core burning phase, the temperature is high enough (T$_9$ $\sim$ 0.3) to efficiently activate ^{22}Ne$(\alpha,$n$)^{25}$Mg. The reaction ^{22}Ne$(\alpha,$n$)^{25}$Mg is the main neutron source for the s-process in massive stars.
As convective core helium burning proceeds, the isotopes ^{12}C, ^{16}O, ^{20,22}Ne and ^{25,26}Mg become the most abundant. Carbon burning ignites in the core and in a convective shell at T$_9$ $\sim$ 1. The energy release is mainly governed by the reactions :

$$^{12}C + ^{12}C \rightarrow ^{24}Mg^* \rightarrow ^{20}Ne + ^4He \qquad (2.11)$$
$$\rightarrow ^{24}Mg^* \rightarrow ^{23}Na + p \qquad (2.12)$$
$$\rightarrow ^{24}Mg^* \rightarrow ^{23}Mg + n \qquad (2.13)$$

With the release of α particles through the ^{12}C$(^{12}$C,$\alpha)^{20}$Ne reaction, α capture reactions are activated and proceed over the ashes of the previous helium core. The strongest α capture reactions are ^{16}O$(\alpha,\gamma)^{20}$Ne, ^{20}Ne$(\alpha,\gamma)^{24}$Mg and ^{22}Ne$(\alpha,$n$)^{25}$Mg. With the carbon burning shell exhausting its ^{12}C, the isotopes ^{16}O, ^{20}Ne, ^{23}Na and ^{24}Mg become the most abundant leading up to the next burning stage. At the end of carbon burning, at solar metallicities, 50% of ^{22}Ne still remains[7].
The next burning stage, Neon burning (T$_9$ $\simeq$ 1.5), ignites after the central carbon exhaustion and is dominated by the reactions ^{20}Ne$(\gamma,\alpha)^{16}$O and ^{20}Ne$(\alpha,\gamma)^{24}$Mg.
At this point, strong neutron densities ($\sim$ 10^{15} cm^{-3}) are produced by the reactions ^{25}Mg$(\alpha,$n$)^{28}$Si and ^{26}Mg$(\alpha,$n$)^{29}$Si[6, 26].

2.2 Stellar Sites for the S-Process

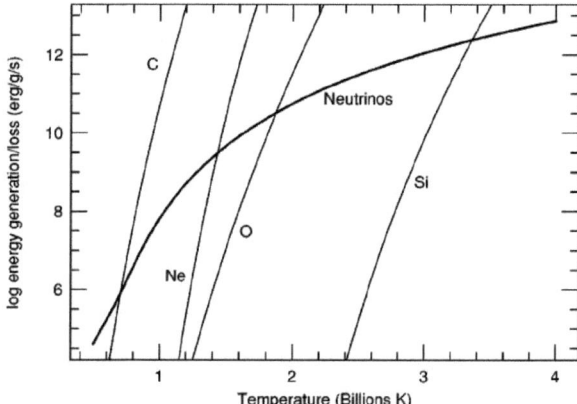

Figure 2.4: Energy generation of the advanced burning stages of a massive star relative to the burning temperature. The dark line labeled "Neutrinos" represents the neutrino losses as a function of temperature[10].

Further burning stages such as oxygen burning are ignited subsequently until the life of the star ends with the iron core collapse and the following supernova explosion[10].
As previously noted, the reaction ^{22}Ne$(\alpha,n)^{25}$Mg is the main neutron source in massive stars. Most of the neutrons released from the reaction ^{22}Ne$(\alpha,n)^{25}$Mg are captured by light nuclei, described as neutron poisons. While the main part of the neutrons is captured in the helium core and the carbon shell, a smaller fraction ($\sim 20\%$) may be captured by Fe seeds[23].
Particulary the recycling effect plays an important role, in which the released neutrons are captured by light poisons but then later on are re-released. The main recycling point involves the reaction cycles : ^{12}C$(n,\gamma)^{13}$C$(\alpha,n)^{16}$O and ^{16}O$(n,\gamma)^{17}$O$(\alpha,n)^{20}$Ne[7].
For example, the competition of the reaction channels ^{18}O$(\alpha,n)^{21}$Ne and ^{18}O$(\alpha,\gamma)^{22}$Ne plays a crucial role in determining the abundance of the Ne isotopes and the previously mentioned neutron source ^{22}Ne$(\alpha,n)^{25}$Mg.
Any uncertainties in the determination of these competing reactions would result in an inaccurate description of the resulting neutron fluxes and isotopic abundances.

2.3 Nuclear Physics behind Nucleosynthesis

To be able to judge the amount of nuclear material synthesized within a certain time span e.g. a burning phase (see equation 2.3), one needs to determine the reaction rates of the reactions involved. The reaction rates are heavily dependent on the probability that the reactions occur, i.e., the cross section σ.

By envisioning a projectile impinging on a target nucleus, one can describe σ, as the probability to cause a well-defined reaction resulting in the release of reaction products and energy. Moving from the classical to the quantum mechanical description of the nuclei, one has to take into consideration such properties as nuclear charge, angular momentum, projectile energy etc.

The cross sections of interest are in general energy, and therefore velocity, dependent. Since the nuclear reactions of interest take place within stellar environments, it is helpful to define the reaction rate of a specified reaction as the product of the incoming particle flux $J = N_x v$ and effective reaction area $F = \sigma(v) N_y$:

$$r = N_x N_y v \sigma(v) \qquad (2.14)$$

where N_x (and N_y) define the number of particles per cubic centimeter. To characterize the movement of the particles within the stellar gas, the product σv can be folded with the velocity distribution $\int_0^\infty \phi(v) dv = 1$:

$$\langle \sigma v \rangle = \int_0^\infty \phi(v) \sigma(v) v \, dv \qquad (2.15)$$

which then allows to define the total reaction rate as :

$$r = N_x N_y \langle \sigma v \rangle \qquad (2.16)$$

The stellar gas can be considered as in thermodynamic equilibrium following the stability criteria and energy conservation for a star. Therefore, the velocity distribution of the particles can be described by a Maxwell-Boltzmann distribution. As the reaction rate includes both interacting particles, it can be rewritten with the velocity distributions for both particles :

$$\langle \sigma v \rangle = \int_0^\infty \int_0^\infty \phi(v_x) \phi(v_y) \sigma(v) v \, dv_x dv_y \qquad (2.17)$$

Rewriting v_x and v_y in terms of the relative velocity and the center of mass velocity V, one needs to use the total mass M and the reduced mass μ to describe the reaction rate :

$$\langle \sigma v \rangle = \left(\frac{8}{\pi \mu}\right)^{1/2} \frac{1}{(kT)^{3/2}} \int_0^\infty \sigma(E) E e^{-\frac{E}{kT}} dE \qquad (2.18)$$

Equation 2.18 shows, that the reaction rate is dependent not only on the cross section of the considered reaction but also on the stellar temperature. Therefore, the reaction rate has to be calculated for ranges of temperatures, since the temperature changes as the star evolves[27].

2.3.1 The Astrophysical S-Factor

The investigation of (α,n) reactions requires the consideration of charged particle reactions first. As a positively charged α-particle is moving towards the target nucleus, it will need to overcome the repulsive force known as the Columb barrier or penetrate it. Gamow was able to show that a particle with an energy lower than the the Coulomb barrier, would be able to tunnel through the potential[28]. The particle then interacts with the nucleus and causes a nuclear reaction. This process is occuring with a given probability P :

$$P = \frac{|\Psi(R_n)|^2}{|\Psi(R_c)|^2} \quad (2.19)$$

where R_n represents the nuclear radius, R_c the classical turning point due to the Coulomb barrier and Ψ the wave function. For $R_c \gg R_n$ one can estimate the probability, also known as the Gamow factor :

$$P = exp(-2\pi\eta) \quad (2.20)$$

where η is the Sommerfeld parameter :

$$\eta = \frac{Z_1 Z_2 e^2}{\hbar v} \quad (2.21)$$

The cross section of charged particle reactions is not only dependent on the de Broglie wavelength :

$$\sigma \propto \pi\lambda \propto \frac{1}{E} \quad (2.22)$$

but also heavily dependent on the height of the Coulomb barrier :

$$\sigma \propto exp(-2\pi\eta) \quad (2.23)$$

The astrophysical S-factor S(E) is defined as :

$$\sigma = \frac{1}{E} exp(-2\pi\eta) S(E) \quad (2.24)$$

which varies smoothly for non-resonant reactions and is considered as one of the main characteristic quantities in Nuclear Astrophysics.
Using equation 2.24, one can rewrite the reaction rate in equation 2.18 :

$$\langle \sigma v \rangle = \left(\frac{8}{\pi\mu}\right)^{1/2} \frac{1}{(kT)^{3/2}} \int_0^\infty S(E) exp\left[-\frac{E}{kT} - \frac{b}{E^{1/2}}\right] dE \quad (2.25)$$

with the barrier penetrability b given by :

$$b = (2\mu)^{1/2} \pi e^2 Z_1 Z_2 / \hbar \quad (2.26)$$

For non-resonant reactions equation 2.25 is dominated by the behaviour of the exponential term. By multiplying both terms of the exponential function in the integrand, one can show that the integrand leads to a peak close to an energy E_0, the so called effective burning energy. The peak is known as Gamow peak, while the energy window underneath is known as Gamow window. Considering that the S-factor is almost constant over small energy windows, one can extract the S-factor out of the integrand in equation 2.25 :

$$\langle \sigma v \rangle = \left(\frac{8}{\pi\mu}\right)^{1/2} \frac{1}{(kT)^{3/2}} S(E_0) \int_0^\infty exp\left[-\frac{E}{kT} - \frac{b}{E^{1/2}}\right] dE \quad (2.27)$$

The effective mean energy E$_0$ is then described by the maximum of the integrand in equation 2.27 :

$$E_0 = \left(\frac{bkT}{2}\right)^{2/3} = 1.22(Z_1^2 Z_2^2 \mu T_6^2)^{1/3} keV \qquad (2.28)$$

where T$_6$ is the temperature in 10^6 Kelvin. In general the effective mean energy E$_0$ is too low for direct experimental measurements of the cross section σ which required in the past the extrapolation of experimental data sets measured at higher energies to the energy E$_0$ (see also table 2.1).

Reaction	T$_9$ [K]	E$_0$ [keV]
^{18}O(α,n)^{21}Ne	0.5	724.53
	1	1150.12
	1.5	1507.08
^{25}Mg(α,n)^{28}Si	0.5	966.08
	1	1533.56
	1.5	2009.53
^{26}Mg(α,n)^{29}Si	0.5	967.80
	1	1536.28
	1.5	2013.1

Table 2.1: List of effective burning energies for ^{18}O(α,n)^{21}Ne, ^{25}Mg(α,n)^{28}Si and ^{26}Mg(α,n)^{29}Si. T$_9$ is the temperature in 10^9 K.

The equations above do not take into account that incident particles are able to form an excited state of the compound nucleus at an energy E$_r$. These can exhibit much higher cross sections than the non-resonant part of the reaction.
By taking into account the angular momentum J and the partial widths Γ, the Breit-Wigner formula displays the cross section for a single-level resonance :

$$\sigma_{BW}(E) = \pi\lambda \frac{2J+1}{(2J_1+1)(2J_2+1)}(1+\delta_{12})\frac{\Gamma_a \Gamma_b}{(E-E_R)^2 + (\Gamma/2)^2} \qquad (2.29)$$

Replacing the non-resonant cross section in equation 2.18 by σ_{BW} and performing the actual integration leads to the stellar reaction rate for a narrow resonance :

$$\langle \sigma v \rangle = \left(\frac{2\pi}{\mu kT}\right)^{3/2} \hbar^2 (\omega\gamma)_R exp\left(-\frac{E_R}{kT}\right) f \qquad (2.30)$$

with $\omega\gamma$ as the resonance strength and f the electron screening factor. The electron screening factor describes the shielding effect caused by the free electrons which are surrounding the nuclei[29].
Equation 2.30 is only valid for the assumption that a narrow resonance, Γ/E_R <10%, is at hand. For broad resonances, $\Gamma/E_R \geq 10\%$ the cross section σ has to be rewritten :

$$\sigma(E) = \sigma_R \frac{E_R}{E} \frac{\Gamma_a(E)}{\Gamma_a(E_R)} \frac{\Gamma_b(E)}{\Gamma_b(E_R)} \frac{(\Gamma_R/2)^2}{(E-E_R)^2 + [\Gamma(E)/2]^2} \qquad (2.31)$$

whereas one also has to take into consideration intereference effects while calculating the S-factor and later on the reaction rate[27].

2.4 Reaction Networks

The use of nuclear reaction networks in stellar models allows to properly simulate the nucleosynthesis of isotopes. Precise experimental reaction rates are the key for accurate nucleosynthesis calculations. A reaction network is given by a list of isotopes properly linked by a complete set of nuclear reaction rates (see section 2.1).
The nucleus of the isotope i undergoes four different types of nuclear reactions :

- the destruction of the nucleus via a two-body reaction
- the production of the nucleus via a two-body reaction
- β^--decay where the nucleus is the daughter and therefore being produced
- β^--decay where the nucleus is the mother and therefore being destroyed

With Y_i defined as the molar fraction of an isotope i, ρ the density and N_A the Avogadro number, one can establish a differential equation to describe the evolution over a certain time t :

$$\frac{dY_i}{dt} = \rho N_A \left(-\sum_j Y_i Y_j \langle \sigma v \rangle_{ij} + \sum_l Y_l Y_k \langle \sigma v \rangle_{lk} - Y_i \lambda_i + Y_m \lambda_m \right) \quad (2.32)$$

where each term respectively represents the above mentioned nucleosynthesis steps. In equation 2.32 three body reactions, etc. are not included.
In this work the NUGRID PPN post-processing code will be employed, which has been developed by Herwig et al.[5, 30, 31]. The post-processing code relies on previous stellar model calculations for the basic structure of a stellar scenario. It allows to calculate complete isotopic abundances for different stages of the stellar evolution.
Essential is the use of most accurate experimental data for the use of the reactions rates. The achieved results by these computational calculations can then be compared with observational evidence and give indications about the accuracy of the stellar models and experimental data sets used.

2.5 Observational Evidence - Meteoritic Grains

Measurements in presolar grains are the most powerful constraint for stellar nucleosynthesis. They amend astronomical observations of stars by extending to measurements of isotopes that are not available to astronomical observations.

While formerly it was believed that all presolar material was vaporized in a hot solar nebula and that all primordial material was isotopically homogenized before forming the solar system, the discovery of isotopic anomalies in small portions (grains, ppm-level) of meteoritic material confirmed a different origin of at least some of the material (see table 2.2) [32, 33, 34].

A presolar grain has the isotopic composition of the stellar atmosphere from which it condensed. This composition is governed by (a) the galactic history of the material from which the star formed, (b) by nucleosynthetic processes in the star and (c) mixing processes in which synthesized material is dredged up into the envelope of the star (see section 2.2). The formation of the presolar grain takes place once the temperatures of expanding envelopes or supernova ejecta is low enough for the condensation of minerals.

Additionally, astronomical observations complement the information from the isotopic compositions of presolar grains assigned to certain stellar scenarios. For example, SiC grains were assigned to originate mostly from AGB stars by comparing the $^{12}C/^{13}C$ ratios and emission features of SiC (see table 2.2) [35]. With the help of measurements on presolar grains one can constrain stellar models even to the point of specific behaviours such as convective mixing of envelopes etc. This allows not only to compare the nucleosynthesis of heavy-mass nuclei by neutron capture, but also the nucleosynthesis of lower mass nuclei by charged-particle reactions.

A fundamental validation of experimental reaction rates and calculated isotopic abundances is therefore provided by the isotopic composition of presolar grains. This validation will be elaborated on recently measured presolar grains and experimental results achieved within this thesis.

Grain type	Size	Abundance	Stellar sources
Diamond	2nm	1400 ppm	SN?
Silicon carbide	0.1-20 μm	15 ppm	AGB, SNe, J-stars, Novae
Graphite	1-20 μm	1-2 ppm	SNe, AGB
Silicates in IDPs	0.2-1 μm	>375 ppm	RG, AGB, SNe
Silicates in meteorites	0.2-0.9 μm	>180 ppm	RG, AGB, SNe
Oxides	0.15-3 μm	>100 ppm	RG, AGB, SNe
Silicon nitride	0.3-1 μm	$\sim$ 3 ppb	SNe
Ti-,Fe-,Zr-,Mo-carbides	10-200 nm		AGB, SNe
Kamacite, Iron	$\sim$ 10-20nm		SNe

Table 2.2: Types of presolar grains in primitive meteorites and interstellar dust particles (IDPs) from [35].

Chapter 3

Previous Results

As mentioned previously (see section 2.2.2), the reactions ^{25}Mg(α,n)^{28}Si and ^{26}Mg(α,n)^{29}Si become neutron sources during the neon burning phase in massive stars. Additionally, the competition between the reactions ^{18}O(α,n)^{21}Ne and ^{18}O(α,γ)^{22}Ne is crucial for the recycling effect and the synthesis of the neon isotopes. The ideal situation would be to have reliable experimental data at stellar temperatures to accurately judge the impact of those reactions.

Calculating the effective mean energy for these reactions following equation 2.28 at different temperatures (see also table 2.1), leads to the conclusion that the reactions have to be measured accurately below a laboratory energy of 1500 keV. In this chapter it will be shown that sufficient experimental data in this energy range has not been available to date.

3.1 ^{25}Mg(α,n)^{28}Si & ^{26}Mg(α,n)^{29}Si

In 1962, Bair et al. were the first to report successful measurements on ^{26}Mg(α,n)^{29}Si[36]. Unfortunately, their measurements were only performed down to a laboratory energy of 3.1 MeV which is not sufficient for nucleosynthesis purposes. Their measurements have an error of up to 50%, due to the unknown contaminations of their targets. They were especially concerned with the ratio between MgO and Mg in their targets, since they evaporated MgO onto a tantalum backing and were not able to determine the Mg:MgO ratio and the target thickness.

Russell et al. used a similiar technique as Bair et al. and extended the energy range for their measurements down to 2.5 MeV[37]. For the detection of the promptly released neutrons a BF$_3$ counter was used, which did not allow a direct separation of neutrons from the reaction ^{26}Mg(α,n)^{29}Si or neutrons being released from other reactions such as ^{13}C(α,n)^{16}O. The quality of spectroscopic information from γ-spectra was further diminished by the occurence of the ^{13}C(α,n)^{16}O reaction as well an the use of NaI(Tl) detectors. Russell and his colleagues used evaporated targets in which MgO was mixed with zirconium. It was believed that zirconium reduced MgO and that the resulting elemental magnesium evaporated onto copper backings. No information on the possible oxygen, respectively MgO, content of the targets is given.

Namjoshi and Bassey II were the only ones later performing experiments to lower energies after Russell, but not down to energies essential for Nuclear Astrophysics[38, 39]. On the other hand, there seemed to be no strong need to investigate the reactions ^{25}Mg(α,n)^{28}Si and ^{26}Mg(α,n)^{29}Si, since ^{13}C(α,n)^{16}O and ^{22}Ne(α,n)^{25}Mg were considered as the major neutron sources for different burning phases in stellar scenarios.

Van de Zwan and colleagues were the first ones to report the cross sections for ^{25}Mg(α,n)^{28}Si down to a laboratory energy of 1.8 MeV using similar techniques as Bair et al.[40]. These authors used different backing materials, but did not mention any contamination of their targets by carbon or oxygen. In addition the use of surface barrier detectors proved to be unsuitable for neutron spectroscopy.

Motivated by the work of Howard et al., Anderson and his colleagues were able to measure both reactions down to an energy of 1.6 MeV and to determine the respective reaction rates[41, 42]. Their goal was to determine whether both reactions could play a role as neutron sources during explosive neon buring at temperatures $T_9 \sim 3$. As detection techniques two methods were employed : The first one was a germanium detector to detect the γ-rays, while a long counter based on BF$_3$ tubes was used to detect the neutrons. This allowed for the correction of additional neutrons being released by the prominent background reactions ^{13}C(α,n)^{16}O and ^{18}O(α,n)^{21}Ne. Evaporated targets were used where MgO was reduced with tantalum powder and then evaporated onto tantalum backings. The thickness of the targets was determined via resonances of the reactions ^{25}Mg(p,γ)^{26}Al and ^{26}Mg(p,γ)^{27}Al.

No information about the physical location of the background reactions is given. A possible contamination of the beamline or the target itself is not taken into consideration. Secondly, energy steps of 50 keV were performed which does not allow the resolution of possible narrow resonances. The use of detection systems with an over all efficiency of a few percent and a reported failure in the detection system contributed further to the uncertainties. Anderson and his colleagues did not compare their experimental results with those of previous authors. They conclude that the reactions ^{25}Mg(α,n)^{28}Si and ^{26}Mg(α,n)^{29}Si could possibly play a role during the process proposed by Howard et al.

3.1 ^{25}Mg(α,n)^{28}Si & ^{26}Mg(α,n)^{29}Si

Küchler & Wieland Measurements

Küchler investigated both reactions utilizing spectroscopy of the released neutrons. For his measurements he used liquid scintillators (Type NE213) at 0, 60 and 90 degrees in order to be able to gain spectroscopic information[43]. Additionally, a ^{3}He ionization chamber was used for a few measurements. Evaporated Mg targets were used which were produced via the reduction of MgO and the following evaporation onto oxygen-free (OFHC) copper. Similar to previous authors no precise measurements of the actual target thickness were performed. No information on the contamination of the targets regarding carbon and oxygen is reported. Since the targets were produced at the MPI für Kernphysik in Heidelberg and then transported to Stuttgart, possible oxidation of the Mg-layer should have been taken into account. In fact, Küchler was able to show that layers of oxygen contributed to his experimental data by comparing them to measurements previously performed by Bair et al.[44]. He also notes that even the use of spectroscopic detectors did not allow the separation of neutrons being released for example by the ^{17}O(α,n)^{20}Ne reaction from those released by the magnesium reactions. Additionally, the author does not report in detail how possible background contributions were subtracted from the experimental data. Additional problems are indicated by the observation that the used targets showed burn and blistering effects after irradiation. The target thickness must have been reduced over time and, if not checked on a regular base, the experimental data were not accurately analyzed.

In spite of this, Küchler calculated the S-factor for both reactions. For ^{26}Mg(α,n)^{29}Si the S-factor was rising for lower energies and for ^{25}Mg(α,n)^{28}Si the S-factor was a smooth function. No information is given on reaction rates and possible influence on stellar nucleosynthesis.

Wieland reanalyzed the experimental data given by Küchler. He came to the conclusion, that the ^{13}C impurities and the according corrections concerning the ^{13}C(α,n)^{16}O reaction were not performed correctly. Implementing a new analysis method and not being able to draw the same conclusion led to the need of new measurements[45].

For these, a neutron detector based on 16 ^{3}He tubes embedded in a 4π polyethylene matrix was used. Another improvement was the use of implanted targets to reduce the impurities and therefore the background. The beamline setup was improved to reduce beam induced background reactions.

The use of a 4π neutron detector with high efficiency resulted in an improvement in yields and therefore reduced beamtime and target destruction. However, no detailed information on the detection efficiency and its determination is given. Wieland only refers to the Monte-Carlo code MCNP and the determination of the efficiency only based on computational calculations. In general, spectroscopy would be only possible if one is able to separate the counting rates in the different ^{3}He tubes and assign them to different neutron energies. This was not possible with the neutron detector available to Wieland.

An area of concern is the lack of information about the used targets. Wieland only notes the implantation parameters of his implanted targets but did not determine the target thickness experimentally. He also used targets previously used by Küchler, but did not discuss possible oxidation of those targets as well.

The usage of gold-plated Cu backings showed improvements in the overall background contribution but remains as a concern. Especially the reported observation of ^{13}C(α,n)^{16}O resonances leads to the conclusion, that the background contribution was still too high for accurate magnesium target measurements.

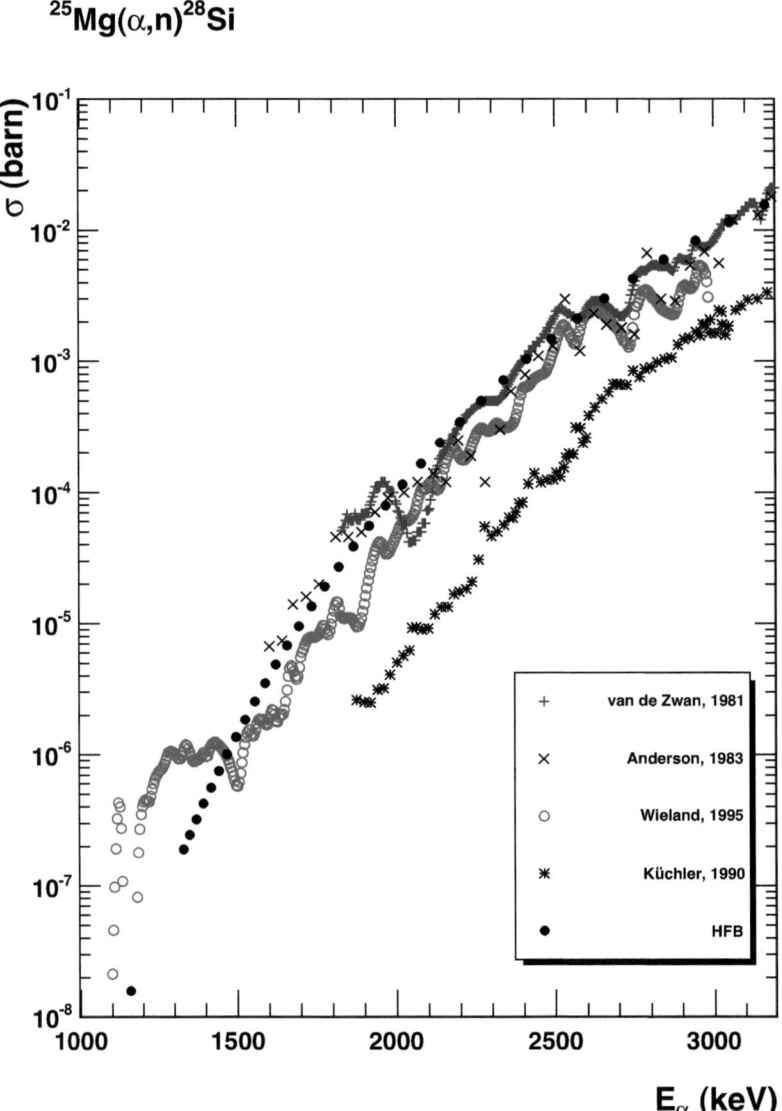

Figure 3.1: Previous results for the total experimental cross section of $^{25}\text{Mg}(\alpha,\text{n})^{28}\text{Si}$ and the results given by the Hauser-Feshbach code (HFB) CIGAR[40, 42, 43, 45, 46].

3.1 ^{25}Mg(α,n)^{28}Si & ^{26}Mg(α,n)^{29}Si

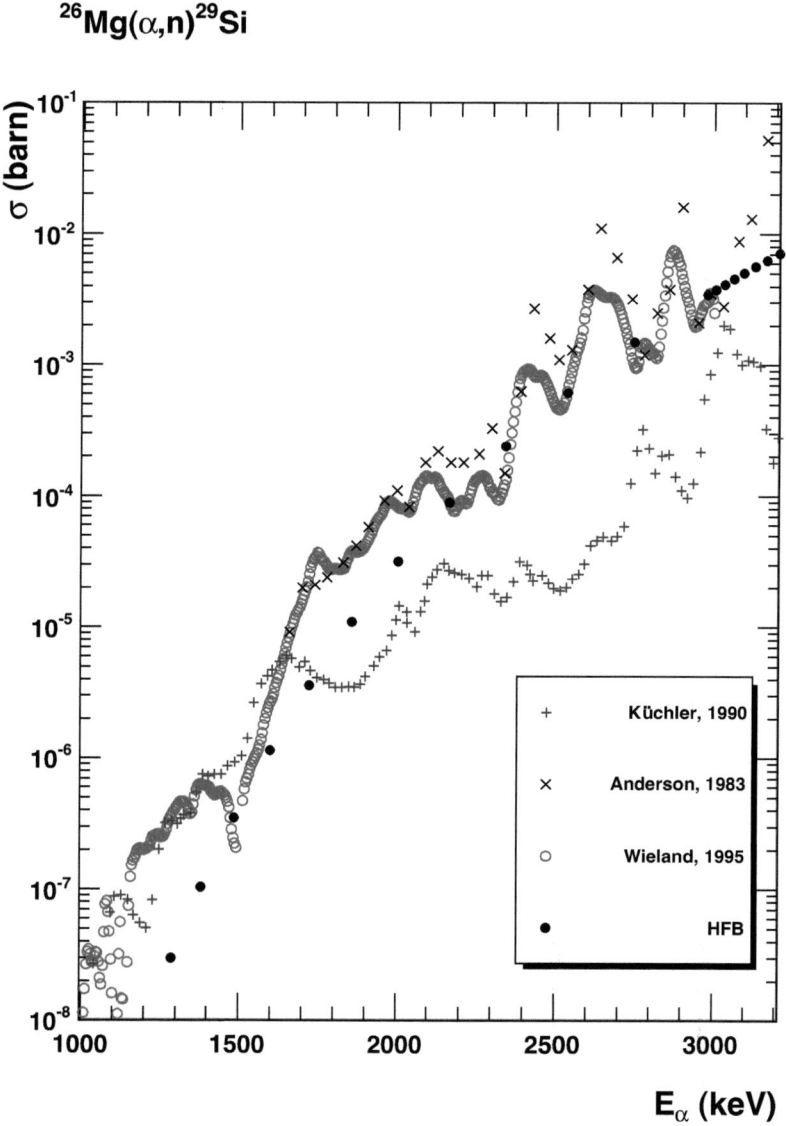

Figure 3.2: Previous results for the total experimental cross section of ^{26}Mg(α,n)^{29}Si and the results given by the Hauser-Feshbach code (HFB) CIGAR[42, 43, 45, 46].

Additionally, the possible contribution of $^{17}O(\alpha,n)^{20}Ne$ and $^{18}O(\alpha,n)^{21}Ne$ due to oxidation is not considered. Probably the lack of accurate experimental data gave no solid foundation for possible arguments toward background contributions by $^{17}O(\alpha,n)^{20}Ne$ and $^{18}O(\alpha,n)^{21}Ne$. With these constraints on his experimental data, Wieland only derived upper limits for the cross section of $^{25}Mg(\alpha,n)^{28}Si$ and $^{26}Mg(\alpha,n)^{29}Si$ below 1.5 MeV.

Wieland also investigated the behaviour of the S-factor and the reaction rate of both reactions, comparing them to the theoretical data given by Woosley and Caughlan [47, 48]. For a temperature range of $T_9 = 0.1$ - 10 the reaction rates differed up to a factor of 5, especially in the low temperature region. Wieland concludes in his diploma thesis that the reactions $^{25}Mg(\alpha,n)^{28}Si$ and $^{26}Mg(\alpha,n)^{29}Si$ have an impact on nucleosynthesis but does not give quantitative arguments.

3.1 ^{25}Mg(α,n)^{28}Si & ^{26}Mg(α,n)^{29}Si

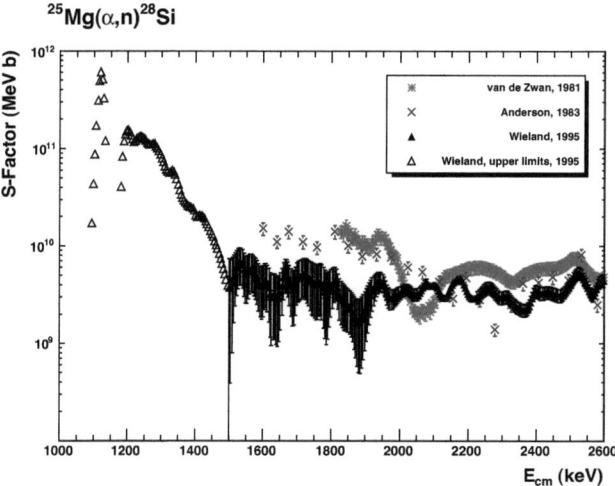

Figure 3.3: Previous results for the astrophysical S-factor of ^{25}Mg(α,n)^{28}Si[40, 42, 45].

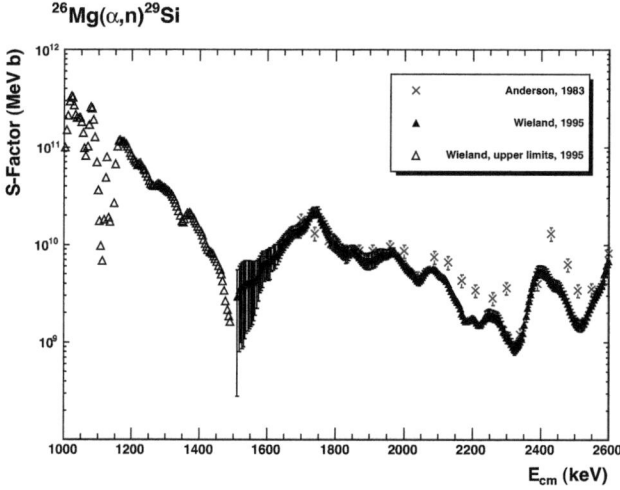

Figure 3.4: Previous results for the astrophysical S-factor of ^{26}Mg(α,n)^{29}Si[42, 45].

Current State of Research

Since the results of Küchler and Wieland were obtained, new theoretical information on the possible role of ^{25}Mg(α,n)^{28}Si and ^{26}Mg(α,n)^{29}Si has become available (e.g. [6, 7]). The improvement of stellar models and nucleosynthesis codes shows a need for more accurate experimental data, as do measurements on presolar grains from meteorites, the results of which indicate inconsistencies with the currrent reaction rates.

To explain anomalous isotopic abundances of the silicon isotopes in main-stream SiC grains, Brown et al. proposed a scenario called magnesium burning[49]. The proposed astrophysical site for magnesium burning are AGB stars with M $\sim$ 6-9M$_\odot$. By artificial adjustments of their model, Brown et al. were able to reproduce the abundance patterns of the Si isotopes found in mainstream SiC grains from AGB stars. In their conlusions they ask for experimental nuclear data to allow a detailed analysis of their model.

As pointed out by Hoppe et al., however, the proposed Mg burning process by Brown et al. and the implied astrophysical model did not agree with measurements on SiC grains [50]. Hoppe notes that the correlation of isotopic abundances of one isotopic chain does not allow to make accurate assumptions upon the astrophysical model. For example, as one adjusts the astrophysical model to reproduce the isotopic ratios of the Si isotopes by introducing ^{25}Mg(α,n)^{28}Si and ^{26}Mg(α,n)^{29}Si as possible sources, one has to take into account higher neutron fluxes as well. These increased neutron fluxes immediately effect the nucleosynthesis of isotopes sensitive to neutron capture nucleosynthesis. By taking into account the abundances of ^{50}Ti and not being able to reproduce them with Brown's model of the Mg burning process, Hoppe showed that such a process is not suitable for single AGB star models.

The conclusion drawn by Hoppe was followed by discussions about the stellar origin of the isotopic ratios of the Si isotopes which has been carrying on until today. As new computational codes and nuclear data became available, the stellar origin was assigned to multiple AGB stars with different metallicities[51, 52, 53].

The discovery of an usual presolar SiC grain of type X (supernova origin) turned the attention of Hoppe and his colleagues on oxygen and neon burning zones based on the stellar models developed by Rauscher et al.[54, 55]. The isotopic abundance ratios predicted by the stellar models were however not reproduced and led to a detailed investigation by these authors. The reactions rates used by Rauscher and adopted by Hoppe et al. are based on the calculations by Fowler et al. and the NACRE collaboration[56, 57].

The reaction rates for ^{25}Mg(α,n)^{28}Si and ^{26}Mg(α,n)^{29}Si of the NACRE collaboration are based upon the experimental data of Wieland and therefore include the inconsistencies regarding the S-factor and the reaction rates discussed above. This leads to reactions rates that are consistently higher than previously obtained values, also acknowledged by Hoppe and collaborators.

Network calculations of O/Si and O/Ne zones involving carbon as well as oxygen burning were performed with modified reaction rates to estimate the possible influence on the Si isotopic abundance distribution. The conclusion of Hoppe et al. is, that the reaction rate of ^{26}Mg(α,n)^{29}Si should be increased by a factor of 2-3 (depending on the stellar model) in order to reproduce the observed abundance patterns.

Hoffman et al. independently carried out network calculations concerning nucleosynthesis in massive stars[6]. One result of their calculations is, that during convective carbon and neon burning the reaction ^{26}Mg(α,n)^{29}Si could lead to the synthesis of nuclei that are usually bypassed by the s-process. Following Hoffman's results, Pignatari was

3.1 ^{25}Mg(α,n)^{28}Si & ^{26}Mg(α,n)^{29}Si

able to illustrate that during convective carbon burning the reactions ^{25}Mg(α,n)^{28}Si and ^{26}Mg(α,n)^{29}Si should not play an important role (see figure 3.6) [7]. Their results rely on the reaction rates achieved by Wieland as well.

As it will be elaborated later, calculations based on the NUGRID PPN network code show similar effects, for example on the distribution of the Ba isotopes.

Both, stellar models and measurements on presolar grains, therefore show a need for more accurate experimental data on the reactions ^{25}Mg(α,n)^{28}Si and ^{26}Mg(α,n)^{29}Si at energies of astrophysical interest.

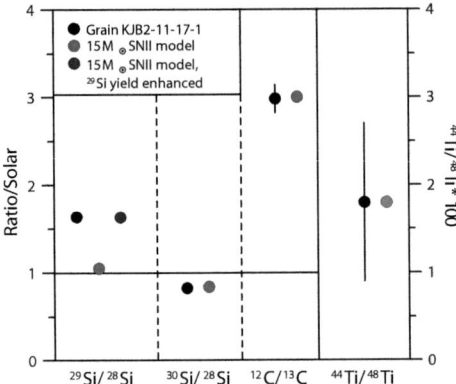

Figure 3.5: Shown are the isotopic ratios of a presolar grain found by Hoppe et al. compared to calculations performed with an enhanced reaction rate of ^{26}Mg(α,n)^{29}Si. The match of the calculations with an artificially enhanced reaction rate with the presolar grain measurements shows the need for an improved experimental data set of ^{26}Mg(α,n)^{29}Si[54].

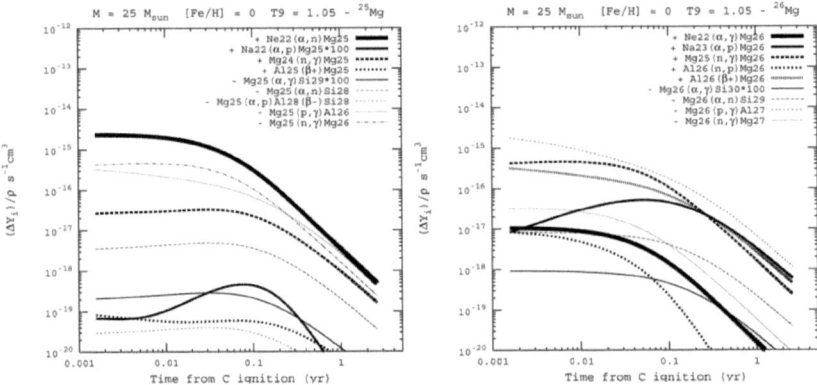

Figure 3.6: Shown is a comparison between the different reaction channels on the isotopes ^{25}Mg and ^{26}Mg based on computational calculations for convective carbon burning[7]. Note, that both reactions do not play a significant role during this burning phase.

3.2 ^{18}O(α,n)^{21}Ne

Bair et al. were the first ones to measure the reaction ^{18}O(α,n)^{21}Ne at energies relevant for nuclear astrophysics[36, 44]. They were able to determine the cross section to an laboratory energy of about 1050 keV. However, their improved experimental techniques (e. g. a gas target system) did not allow an accurate analysis for the purpose of Nuclear Astrophysics. The data of Bair et al. suffered mainly from low resolution, low ^{18}O enrichment and the background contribution from the reaction ^{13}C(α,n)^{16}O.

Denker was able to determine the reaction rate of ^{18}O(α,n)^{21}Ne down to the threshold energy E_{thres} = 850.95 keV[8]. A sophisticated gas target system as well as a highly enriched target gas were used to perform the experimental measurements. Denker was also first in addressing the comparison of the reaction channels ^{18}O(α,γ)^{22}Ne and ^{18}O(α,n)^{21}Ne (section 2.2.2). Until today, the measurements by Denker have not been confirmed by an independent experiment, however.

More recent measurements of the reaction ^{18}O(α,γ)^{22}Ne revealed new details about its reaction rate. Dababeneh et al. were able to evaluate the influence of their experimental results on the reaction rate, but did not draw a comparison to the ^{18}O(α,n)^{21}Ne reaction[58]. Similar to ^{25}Mg(α,n)^{28}Si and ^{26}Mg(α,n)^{29}Si, Pignatari was able to show, that ^{18}O(α,n)^{21}Ne does not play a role during carbon burning as well (see figure 3.7) [7].

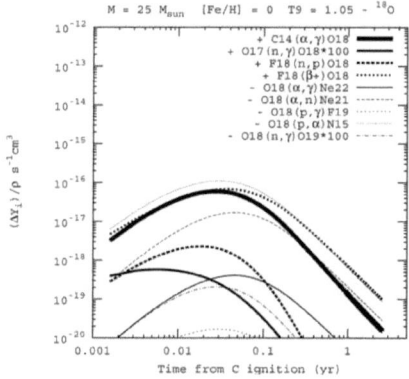

Figure 3.7: Comparison between the different reaction channels on ^{18}O based on computational calculations for convective carbon burning[7]. Similar to the reaction ^{25}Mg(α,n)^{28}Si and ^{26}Mg(α,n)^{29}Si the reaction ^{18}O(α,n)^{21}Ne does not play a major role during convective carbon burning.

Following the comparison of both reaction channels, the nucleosynthesis of the neon isotopes has been investigated in the past as well. Specifically the abundance patterns of the neon isotopes in meteoritic inclusions were not represented accurately with current stellar models (see figure 3.8) [59, 60].

The missing comparison with the most recent results on competing reaction channels, disagreements of current star models with abundance patterns in presolar SiC grains and finally the ambiguity of the experimental measurements reveal also a clear demand for new experimental measurements of the reaction ^{18}O(α,n)^{21}Ne.

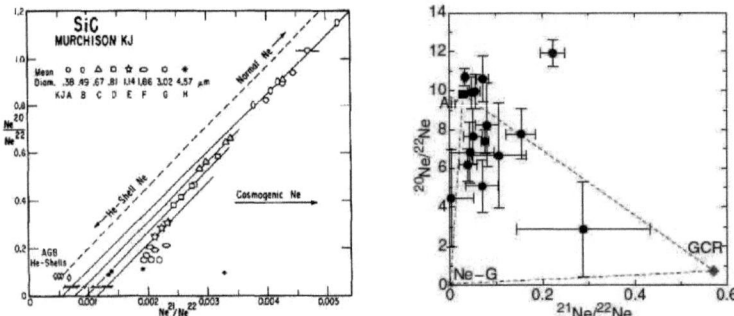

Figure 3.8: Displayed are three isotope plots to compare between measured isotopic abundances of neon in presolar grains and predicted isotopic abundances by stellar models (Ne-G) (left from [59], right from [60]).

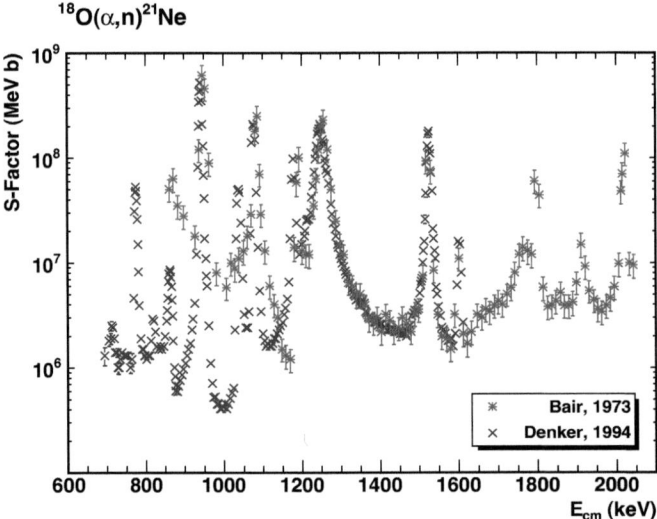

Figure 3.9: Previous results for the astrophysical S-factor of ^{18}O(α,n)^{21}Ne[8, 44].

3.2 ^{18}O(α,n)^{21}Ne

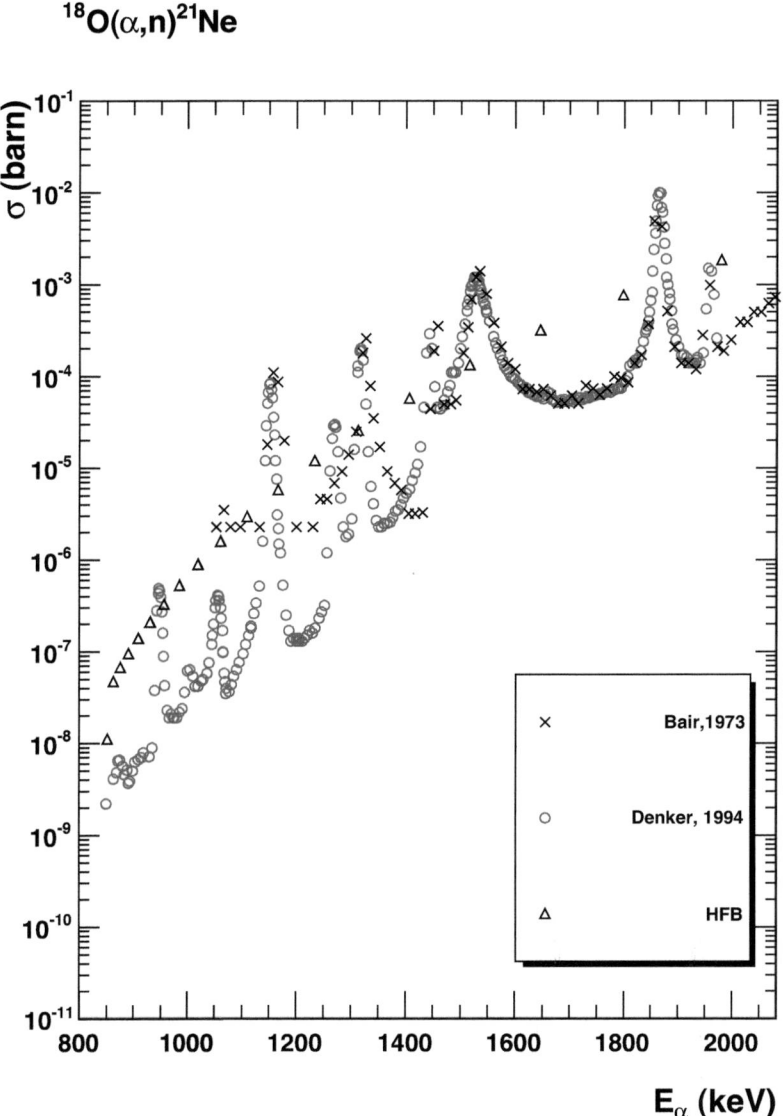

Figure 3.10: Previous results for the total experimental cross section of ^{18}O(α,n)^{21}Ne and the results given by the Hauser-Feshbach code (HFB) CIGAR[8, 36, 44, 46].

Chapter 4

Experimental Techniques and Procedures

Nuclear reactions are verified by the detection of their reaction products. For the case of ^{18}O$(\alpha,n)^{21}$Ne, ^{25}Mg$(\alpha,n)^{28}$Si and ^{26}Mg$(\alpha,n)^{29}$Si, the reaction products are a neutron and an excited nucleus. The neutron emerges from the target while the excited nucleus decays via γ-decay into its ground state.

After reviewing the different possible detector techniques, it was decided to detect the neutrons resulting from the reactions. The advantage detecting the neutrons compared to the γ-rays is a high detection efficiency and a relatively flat efficiency behavior. For the acceleration of the α-particles, the KN Van-de-Graaff accelerator (KN) at the Nuclear Science Laboratory at the University of Notre Dame (NSL) was used. The particle beam was directed onto evaporated and anodized targets. Five experimental beam times were performed, during which accelerator calibration, detector calibration and production runs were executed.

4.1 The KN Accelerator and Beam Transport System

The experiments were exclusively performed at the NSL and the KN accelerator. The KN accelerator is a single-ended Van-de-Graaff accelerator containing an internal RF-tube ion source. The ion source allows to switch between α-beam and proton-beam without opening the tank of the accelerator. The experiments were performed at energies between 700 and 2700 keV.

As the accelerated ions leave the accelerator tank, they enter a high-vacuum, ion optical beam transport system (beam line) which carries and directs the beam to the desired target. The first ion optical element of importance is the analyzing magnet and its slits (analyzing slits) located behind the analyzing magnet. To ensure that the particles impinging on the target have the same energy, they are sent through the analyzing magnet. The field in the magnet is set, so that the particles having the desired energy are redirected while other particles do not pass the analyzing magnet. The analyzing slits ensure, that the terminal voltage of the accelerator is adjusted to the right particle energy. The magnetic field is measured via a NMR probe.

Before each experimental beam time, the magnetic field setting of the analyzing magnet was calibrated with respect to particle energy. This was realized by changing the particle energy until a known resonance of a specific reaction was observed. One can then correlate the resonance energy with the field setting in the analyzing magnet and derive a calibration function for the particle energy. A similar method is the measurement across the threshold of a well known reaction. Both methods were used multiple times and involved the reactions ^{27}Al(p,γ)^{28}Si, ^{51}V(p,n)^{51}Cr and ^{18}O(α,n)^{21}Ne (see table 4.1 and figure 4.1) [27, 61, 62].

Reaction	E_{lab} [keV]	Type of energy calibration	Particle detected
^{27}Al(p,γ)^{28}Si	992	Resonance	γ
^{18}O(α,n)^{21}Ne	1866	Resonance	neutron
^{51}V(p,n)^{51}Cr	1565	Threshold	neutron

Table 4.1: List of reactions used for the KN energy calibration before every experimental beam time.

As the particle beam exits the analyzing magnet, a second magnet (switching magnet) directs the beam. The beam is directed either to a beamline designed for Rutherford-Backscattering (RBS beamline) measurements or to a beamline for measurements of (p,γ) and (α,n) reactions (0° beamline).

4.1.1 RBS Beamline

Rutherford-Backscattering is a technique that utilizes the scattering of incoming particles on a thin film probe (for example a target) to determine its composition.

Based on the evaluations of previous measurements of the ^{25}Mg(α,n)^{28}Si and ^{26}Mg(α,n)^{29}Si reactions, it was decided to measure the composition of the Mg targets with the RBS method. As the α-beam passes through a collimator, it impinges on the RBS target at 0°. Depending on its energy, the α-particle is backscattered and then detected by a Si detector, placed at an angle θ.

4.1 The KN Accelerator and Beam Transport System

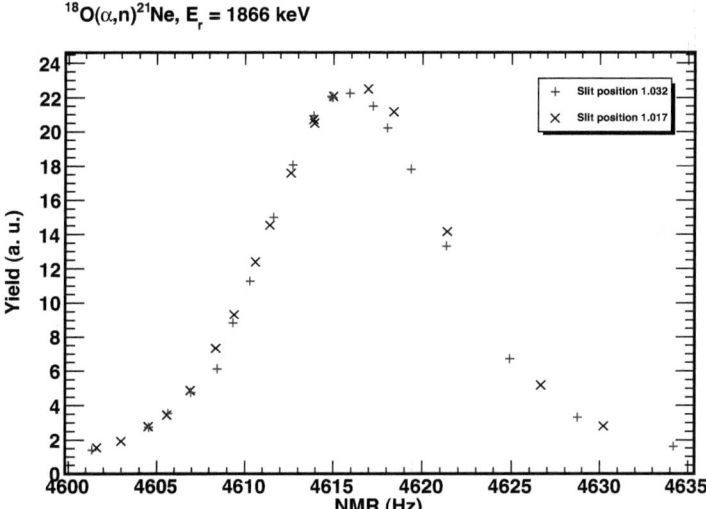

Figure 4.1: Plot of the ^{18}O(α,n)^{21}Ne reaction at E$_\alpha$ = 1866 keV to determine the particle energy calibration at different analyzing slit positions. The different slit positions showed only a marginal effect on the particle energy calibration.

The initial Energy E_i of the incoming particle is proportional to its detection energy E_1 :

$$k = \frac{E_1}{E_i} = \left(\frac{m_1 cos(\theta) \pm \sqrt{m_2^2 - m_1^2 sin(\theta)^2}}{m_1 + m_2} \right) \quad (4.1)$$

The cross section is heavily dependent on the atomic number of the involved nuclei :

$$\sigma_{Rutherford} = \left(\frac{Z_1 Z_2 e^2}{4 E_i} \right)^2 \frac{1}{[sin(\theta/2)]^4} \quad (4.2)$$

From equations 4.1 and 4.2 one can show that different scattering partners in a probe result in different yields and positions in the spectra. When a Mg layer completely oxidizes, MgO is formed which in absolute numbers gives 1 oxygen nucleus per 1 magnesium nucleus. During a RBS measurement this can be illustrated when the yield of oxygen is 2.25 ($12^2/8^2$) times smaller than the magnesium yield. The stoichiometry of the target can then be derived from the RBS measurement.

Since the RBS cross section is heavily dependent on the atomic number, target backings of high atomic number materials are unfavorable. To ensure a clear distinction between the different nuclei in the targets circular carbon disks were placed next to the used target backings during the target production process. Carbon was used, since it is the only material with an atomic number less than oxygen that was practicable to use. As a result, only nuclei with Z>6 can be clearly resolved.

For the Mg targets only the RBS method could be utilized, since the production process of the ^{18}O targets (anodization) did not allow a parallel production of samples useful for the RBS method.

4.1.2 0° Beamline

The experimental measurements on ^{18}O(α,n)^{21}Ne , ^{25}Mg(α,n)^{28}Si and ^{26}Mg(α,n)^{29}Si were performed on the 0° beamline. After the switching magnet, the beam passes through a set of magnetic quadrupoles, electrical steerers and a beam profile monitor until it reaches the target chamber. The ion optical modules were used to ensure a consistent beam profile during the course of the experiments. The beam was also analyzed by slits which were set up along the beam line at different locations. At the target chamber two pairs of slits were mounted, directly followed by a copper tube.

The copper tube ended right before the target itself and was used as a cold trap. A suppression voltage of -700 V was applied additionally and the vacuum of the target chamber was kept at a few 10^{-7} torr.

Since it was not necessary to shut down the accelerator to switch between proton and α beams, the setup of the ion optical system was not changed during the course of the experiments and only required fine tuning. A strict beam tuning routine was developed (see appendix A) to ensure that the beam consistently hit the same area of the target and was rastered over the target regularly. The target was cooled from its backside with a constant flow of deionized water.

4.2 Neutron Detection

Since neutrons do not carry an electrical charge, one has to utilize a nuclear reaction to create a electronically convertable signal. In order to do so, the reaction ^{3}He(n,p)^{3}H was used. The reaction has a relatively high cross section for low neutron energies and therefore enhances the detection efficiency (see figure 4.2).

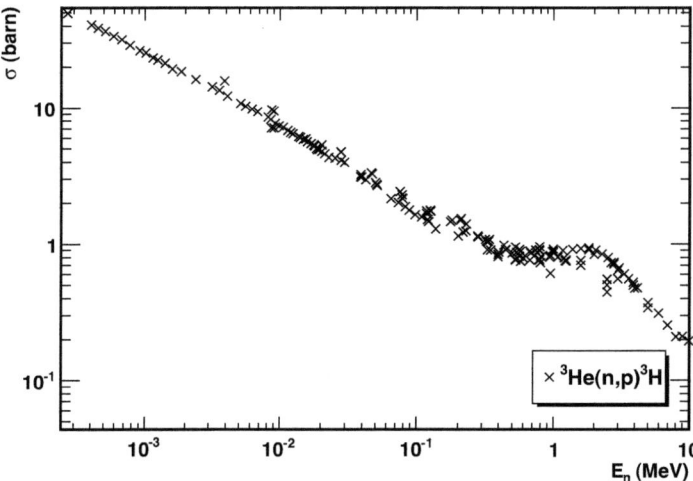

Figure 4.2: Illustrated is the cross section for the reaction ^{3}He(n,p)^{3}H from the ENDF compilation[63].

Based on the first designs by Hanson et al., ^{3}He-proportional counters embedded in polyethylene have been utilized for nuclear physics purposes[64, 65, 66]. 22 ^{3}He proportional counters were lent to the NSL by the University of Mainz.

A single ^{3}He proportional counter consists of an aluminum tube into which highly pressurized ^{3}He-gas is put. In the center of the aluminum tube an anode wire is mounted and put under high voltage. Once a neutron passes the aluminium shielding and interacts with the ^{3}He-gas, the resulting proton causes a discharge which is collected by the anode. The collected charge is then converted into a electronic signal. The basic physical properties of the used ^{3}He counters are shown in appendix B.

Due to the reaction kinematics of the ^{3}He(n,p)^{3}H reaction (Q = 764 keV), the energy of the resulting protons (E_p = 573 keV) and tritons (E_t = 191 keV) is always the same, causing the same signal amplitude for each incident neutron.

Besides a single peak in the spectra one can also observe a tail of the peak, which is caused by the so-called wall effect. The effect is caused by random collisions between the reactions products and the walls of the aluminum tube. As a result, the reaction products lose energy and therefore cause a different signal amplitude.

To determine the optimum operating voltage each single ^{3}He tube was tested with different preamplifiers at different voltages. An operating voltage of 1400 V was found to be optimal.

4.2.1 Moderation of Neutrons

The cross section of the ^{3}He(n,p)^{3}H reaction increases with lower neutron energies. Therefore, one can increase the detection yield by decreasing the neutron energy before it enters the ^{3}He proportional counter. This can be achieved by forcing the neutrons to pass through additional material in which they lose energy (known as moderation or thermalization) before they reach the ^{3}He counter.

Behind the moderation of neutrons stand inelastic scattering processes in a so called moderator material. Ideally the neutrons enter the material, loose energy due to several scattering processes and enter the ^{3}He proportional counter with an energy lower than their original energy[67].

Concerning the scattering processes, materials with a low Z are favored and should not cause other interactions. Polyethylene and similar $(CH)_n$ based materials have been used as moderation materials in the development of neutron detection systems[68, 69, 70, 71]. The advantages are that polyethylene is relatively easy to machine and is not expensive.

Theoretically, the moderation of neutrons can be described by the scattering matrix S(α,β) formalism which is used in simulation frameworks to calculate the moderation process :

$$\sigma(E \rightarrow E', \mu) = \frac{\sigma_b}{2kT}\sqrt{\frac{E'}{E}}S(\alpha,\beta) \quad (4.3)$$

with σ_b being the material-dependent characteristic bound scattering cross sections and μ the cosine of the scattering angle in the laboratory system. The scattering cross section is dependent on momentum (α) and energy (β) transfer :

$$\alpha = \frac{E' + E - 2\sqrt{E'E}\mu}{AkT}, \beta = \frac{E' - E}{kT} \quad (4.4)$$

where A is the ratio of the scatterer mass to the neutron mass.

The moderation method itself introduces the disadvantage that one does not know the initial neutron energy before the neutron enters the moderation material.

4.2.2 Design Principles

To design a neutron detector based on ^{3}He proportional counters and moderation material, one has to follow certain design principles which are crucial for its performance.

The most crucial parameter is the efficiency: the number of detected neutrons per number of emitted neutrons. Since the expected reaction yields are relatively low, the efficiency needs to be maximized and therefore the design of the detector follows primarily this criterion.

A second criterion is that the neutron detector should have a flat response function for different neutron energies and be relatively position insensitive towards the target position. The idea of an easy adaptability of the system has also been incorporated in the design process.

For example, the attachment of additional shielding or moderation material was aimed to be as simple as possible. Another aspect is that the neutron detector should be set up at the end of a shared beamline so that it can be used for other experiments as well.

4.2 Neutron Detection

As a first step in the actual design process, the design of previous neutron detectors and their performances were reviewed. Measurements of low energy reactions in general and delayed neutron emission experiments were major applications for such detection systems[68, 69, 70, 71].

With no natural neutron sources available to deliver neutrons at discrete energies, experimental measurements on nuclear reactions were performed to validate computational simulations and derive an efficiency function later on. To ensure an accurate computational simulation, it was chosen to design a test detector and measure its response. Later on, the experimental results were compared to the simulations. In the simulations one heavily relies on the accuracy of the used scattering and reactions cross sections, the moderation of the neutrons plays a crucial role, since it determines the energy of the neutron when it enters the proportional counter. With knowledge and correlation between experimental performance and computational simulations the final neutron detector was designed and tested.

4.2.3 Computational Simulations

The previously mentioned computational simulations are based on the Monte-Carlo method. The Monte-Carlo method allows to duplicate and model statistical processes such as interactions between particles and matter. The modeling process is based on random number seeds and consists of sequentially modeled processes. For the transport of particles through matter it is crucial that the derived code is able to track the particles and their corresponding interactions. To determine the outcome of each process step the probability functions of each interaction are randomly sampled.

Experimental data sets for different transport and interaction processes and particles are included into the computations to achieve a maximum of accuracy. For additional validation purposes, especially for the physics models used, it was chosen to use two simulation frameworks, GEANT4 and MCNP5[72, 73].

GEANT4

GEANT4 is an object-oriented simulation toolkit primarily delevoped for simulations of experiments at the European Nuclear Research Organisation (CERN). The programming language used is C++ with the code being open source. The toolkit provides the necessary accessories such as geometry design, physical models, materials etc. for the user to be able to simulate a full experimental setup.

The first choice one has to make is to choose the right physical model for the description of the physical interactions. In order to do so, the interaction between neutrons and ^{3}He gas was simulated with different models and also compared with experimental data sets. The neutron energies of interest are below 10 MeV, therefore computational simulations were only performed for initial neutron energies below 10 MeV.

It was found that different physical models incorporated in the computational simulations delivered (for the same geometry) the same results within less than a percent of deviation. The so called QGSP (Quark-Gluon-String-Plasma) model was finally used, since it is reported to produce the best results of thermal neutron scattering in matter[74].

GEANT4 uses also experimental data to model the scattering of thermal neutrons in matter. The experimental data is derived from evaluated data libraries such as ENDF and JENDL[63, 75]. To ensure a correct handling of the modeling and use of experimental data, simple simulations were carried out in which neutrons were directly sent through a large volume of ^{3}He gas. The results were then compared to the evaluated data sets of ENDF/B-VI.8 and JENDL3.3.

To achieve a physical meaningful result, the number of recorded reactions must be weighted by the energy loss of the neutron from its starting point to its interaction point with the gas.

In other terms : Two neutrons that have the same initial energy are not necessarily detected with the same probability, since e.g. scattering causes them to lose different amounts of energy until they reach their interaction point. At their interaction point they will have different energies with which they interact with the gas. Therefore, the relevant cross-sections and detection probabilities will be different.

As a consequence, the results were weighted with respect to the ratio of initial energy and the energy at interaction point. Very good agreement between for example the ENDF data set, the data set used by GEANT4 and the actual calculated data was achieved (see figure 4.3).

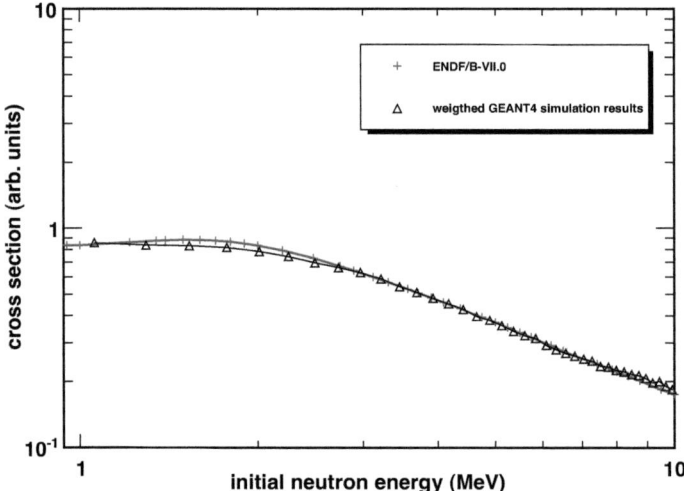

Figure 4.3: Comparison between the experimental data set provided by ENDF and calculated by GEANT4. The GEANT4 results were weighted with respect to the ratio of initial neutron energy and neutron energy at interaction point.

GEANT4 allows to include a parameter (*G4NEUTRONHP_SKIP_MISSING_ISOTOPES*) where the modeling of neutron data sets for isotopes not implemented is skipped. Tests showed, that it had no effect on the results of the experimental setup.

To implement different physical models for the same process, the computation of interactions between particles is based on the calculation of the distance from the origin of the particle to the point of interaction (or decay). The probability of the particle not

4.2 Neutron Detection

interacting with anything within a given distance ℓ is defined as :

$$P(\ell) = e^{-n_\lambda}, \ n_\lambda = \int_0^\ell \frac{d\ell}{\lambda(\ell)} \tag{4.5}$$

where the mean free path λ for interactions in a material of density ρ and the cross section σ are defined as:

$$\frac{1}{\lambda} = \rho \sum_i \frac{x_i \sigma_i}{m_i} \tag{4.6}$$

with i denoting the different isotopes in the material.
This implies that e.g. exact material densities have to be taken into account in order to achieve an accurate experimental description. For example, the ^{3}He-gas in the simulated proportional counter has to be described at 10 atm pressure, since the interaction probability between neutron and the ^{3}He-gas is pressure dependent. Note, that the probability *distribution* can be written as :

$$n_\lambda = -ln(\eta) \tag{4.7}$$

where η is a random number between 0 and 1. Therefore the probability distribution is independent of energy and material.
Despite a precise description of the experimental geometry, it is also necessary to define conditions and qualitative measures for the desired quantity. As previously mentioned, the number of detected neutrons per number of emitted neutrons (the efficiency) for each initial neutron energy is the key parameter. The number of detected neutrons is equal to the number of ^{3}He(n,p)^{3}H reactions within the ^{3}He proportional counters.
The neutrons which enter the ^{3}He proportional counters from the surrounding material were registered. If the neutron interacts with the ^{3}He gas resulting in a proton and a triton, the event would be registered and recorded. Finally, this was defined within the tracking routines of GEANT4 to ensure a precise count of recorded ^{3}He(n,p)^{3}H reactions caused by an incoming neutron.
The simulations were carried out and stored within ROOT histograms to be able to judge the response function of the chosen detector geometry[76]. Tests with different material settings and physics models were performed to ensure a maximum of consistency in the detector performance. For example, it was chosen to vary the density of the moderator material, since polyethylene is available in different densities.
A higher density results in a higher number of scattering partners for the incoming neutron per volume unit. Therefore, an effect on the thermalization process is expected which results in a different detector efficiency. Not only different neutron energies but also parameter studies on different neutron source and target positions were performed. With different source positions it was possible to determine the best target position within the neutron detector and probe it for example against misalignments.
After the parameter studies delivered reliable results, the optimal detector geometry delivered by GEANT4 was implemented into MCNP5 and the results from the two simulations were compared.

MCNP5

As a second simulation toolkit the Monte Carlo N-Particle code (MCNP5) provided by the Diagnostics Application Group at the Los Alamos National Laboratory was chosen[73]. MCNP is based on the programming language FORTRAN and in contrast to GEANT4 not

openly available. Since it is not openly distributed, it is more strict on the user concerning the use of physics models and other capabilities. The definition of experimental setups, event biasing and output definition is completely different from GEANT4.
For the purpose of this thesis it was particularly interesting to make use of the S(α,β) tables in MCNP which are crucial for low neutron energies. MCNP, similar to GEANT4, uses data libraries such as ENDF and the treatment of particle tracks etc. is very similiar. MCNP5 utilizes so called tallies within the calculated scenarios. Tallies are quantities defined in MCNP to deliver an useful output for the user. The restriction on tallies is that they are based on fluences and products or summations in the cross-section libraries. In general, tallies and their functions are dependent on time and energy, normalized per starting particle.

Despite the absolute result given by MCNP5, a relative error is given in the output to allow the user to assign the results a confidence levels. The estimated relative error given by MCNP5 is a confidence statement referring only to the precision of the calculations and not the accuracy.

Tallies were defined which reflect the number of ^{3}He(n,p)^{3}H reactions within the ^{3}He proportional chambers. The tally used is the so-called F4 tally, which is the track length estimate for a flux within a well-defined physical volume :

$$\overline{\phi}_V = \frac{1}{V} \int dE \int dV \int dt \int d\Omega \; \Psi(\vec{r}, \hat{\Omega}, E, t) \tag{4.8}$$

with Ψ being the angular flux given by the product of particle density n and velocity v :

$$\Psi(\vec{r}, \hat{\Omega}, E, t) = vn(\vec{r}, \hat{\Omega}, E, t) \tag{4.9}$$

The average particle flux then becomes :

$$\overline{\phi}_V = \frac{1}{V} \int dE \int dV \int dt \int d\Omega \; vn(\vec{r}, \hat{\Omega}, E, t) \tag{4.10}$$

$$= \frac{1}{V} \int dE \int dV \int dt \; vN(\vec{r}, E, t) \tag{4.11}$$

$$= \frac{1}{V} \int dE \int dV \int ds \; N(\vec{r}, E, t) \tag{4.12}$$

by noting that N($\vec{r}$, E, t) is the density of particles while ds is the differential unit of the track length vdt.

By multiplying the average fluence from equation 4.10 with the cross section of ^{3}He and a normalization factor the tally then represents the number of produced ^{3}H particles as a result of the ^{3}He(n,p)^{3}H reaction. Normalized over the number of initial neutrons, one obtains the efficiency. Therefore the results given by MCNP5 are simple to compare with the GEANT4 results.

4.2.4 Parameter Studies

It was chosen to study the setup of a relatively small test detector before investigating the setup of a full detector. ^{3}He proportional counters were embedded into a block of polyethylene with a source of neutrons in the center. The number of ^{3}He proportional counters was varied between 4 and 8, with the number of detected neutrons increasing with the number the number of counters.

4.2 Neutron Detection

The distance between the ^{3}He counters and the neutron source has a large influence. The greater the distance between the neutron source and the ^{3}He counter, the larger the physical volume of moderation material a neutron has to travel through before interacting with the ^{3}He gas. Hence, the more scattering reactions will take place and accordingly the neutrons will be more strongly moderated.

If the volume of the moderator material between counter and neutron source is further increased, the volume of the moderator will eventually reach a value, at which the moderating properties of the polyethylene turn into shielding properties. The thermalization of the neutrons is then strong enough to cause full energy loss of the neutrons, so they are not able to reach the ^{3}He counters.

Another factor that enters is the density of the moderation material. Polyethylene is available in different densities which causes different amounts of scattering reactions per cm^3. For the design purpose so called high-density polyethylene (HDPE) was implemented. Compared to ultra high molecular weight polyethylene (UHMWPE), HDPE has little branching. In other terms : The polymer chains do split out less into multiple chains as compared to other polyethylene forms. This results in stronger intermolecular forces and tensile strength. HDPE has the advantage that the higher density (0.965 g/cm^3) increases the number of scattering partners for passing neutrons per volume unit. Therefore the moderation abilities should be improved compared to other polyethylene varieties.

Not only the distance between proportional counter and neutron source was found to be crucial, but also the overall amount of polyethylene surrounding the proportional counters. For example, a neutron is able to scatter past a proportional counter and then scatter back toward the proportional counter. By increasing the overall size of the polyethylene volume, also called polyethylene matrix, the efficiency of the overall detection system was increased (see table 4.2).

Size HDPE matrix [cm]	absolute efficiency [%]
15.24 x 15.24 x 15.24	26.15
200 x 200 x 200	39.43

Table 4.2: Comparison of the detection efficiency for differently sized HDPE matrices. The absolute efficiency is the overall efficiency for an energy range from 0 - 6 MeV.

However, limiting factors are the availability, machinable size of polyethylene and the feasibility regarding the physical dimensions of the target chamber etc.

Parameter variations were tracked where one parameter at a time was tracked to allow identification of the optimal configuration. The first parameter varied was the number of ^{3}He counters in one ring arranged around a target holder and embedded in polyethylene. After that, the radius of the ring, i.e. the distance between proportional counters and target, was varied. Following this, the size of the polyethylene matrix was varied until the maximum efficiency was found.

As a last parameter the shielding effect toward neutrons coming from the outside of the detector, for example caused by cosmic radiation, was investigated. Since borated polyethylene is very similar to high density polyethylene, it was chosen to investigate the shielding effects caused by an additional layer of 5 percent borated polyethylene. An additional layer of 2 inch borated polyethylene showed a reduction of neutrons, impinging on the detector from outside, by up to two orders of magnitude.

A property which has not been mentioned so far is the rather low sensitivity toward the

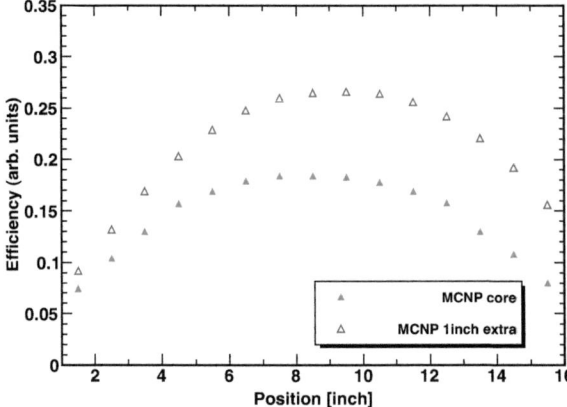

Figure 4.4: Illustrated is the efficiency of the MCNP results for the test detector efficiency (core) and a extra layer of HDPE of 1 inch thickness surrounding the core. The different positions are equal to one inch. Position 8 marks the target being in the center of the detector.

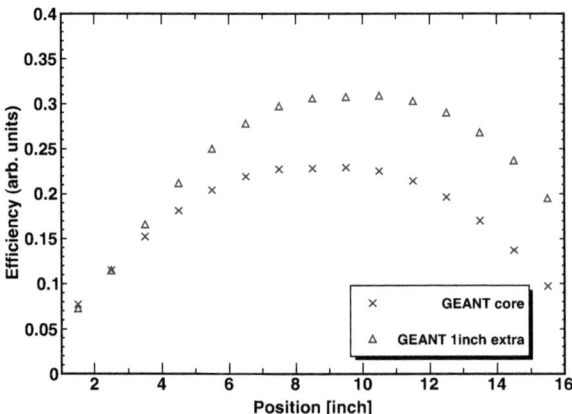

Figure 4.5: Illustrated is the efficiency of the GEANT results for the test detector efficiency (core) and a extra layer of HDPE of 1 inch thickness surrounding the core. The positions etc. are analog to figure 4.4. The difference between the results provided by MCNP and GEANT at e.g. the positions 1 and 2 indicate a different treatment of the neutron moderation.

4.2 Neutron Detection

target position relative to the ^{3}He proportional counters along the beamline axis. The calculations showed that the target position can be off center by almost more than an inch without effecting the detector performance (see figures 4.4 and 4.5).

4.2.5 Design of the Test Detector

A single ring test detector was designed which allowed the attachment of additional polyethylene layers. This feature allowed to verify several parameters experimentally which were optimized in the simulations. The computational results delivered an optimal configuration with ten ^{3}He counters for a single ring at a distance of 6.1 cm from the target. The absolute efficiency was given at about 40 percent for the test detector, but led to a constructional problem.

Since the ^{3}He proportional counters have a diameter of about one inch, the holes to be drilled for ten ^{3}He counters on a circle of 6.1 cm radius would leave too little material between them. This would cause instabilities and deformations within the material and disturb the detector performance.

As the calculated efficiencies suggested, that for a detector with eight ^{3}He proportional counters the absolute detector efficiency would be at most 2 percent lower than for a detector with ten ^{3}He proportional counters, a setup with 8 ^{3}He counters was chosen (see table 4.3).

Number of ^{3}He tubes	absolute efficiency [%]
6	33.90
8	38.16
10	39.95

Table 4.3: Comparison of the detection efficiency for different number of ^{3}He tubes. The distance to the target for each simulation is 6.1 cm. A detector with ten ^{3}He tubes shows the highest efficiency but is compared to a detector with eight ^{3}He tubes not feasible for machining purposes.

Finally, for eight ^{3}He proportional counters the computations delivered a optimal radius of 5.8 cm for the single ring (38.21 %) as compared to the 38.16% for a radius of 6.1 cm. The test detector then consisted of a eight ^{3}He proportional rings on a distance of 5.8 cm from the center of the polyethylene matrix. The polyethylene matrix itself was 6 x 6 x 13 inches in dimensions (see figure 4.6).

The matrix was built out of four separate small matrices which were hold together by four long, tapped polyethylene rods. Additionally, one inch thick sheets were machined, which could be easily screwed onto the core polyethylene matrix. Those were used to test the effects of polyethylene surrounding the ^{3}He proportional counters. The hole for the target chamber and target holder was drilled completely through the polyethylene matrix to allow easy adjustment of positioning and feedthroughs for the water cooling lines. The face of the detector facing upstream was not completely drilled through. This turned out to be an advantage in terms of counter positioning relative to the target. No shielding materials were used in order to simplify analysis of experimental and computational data.

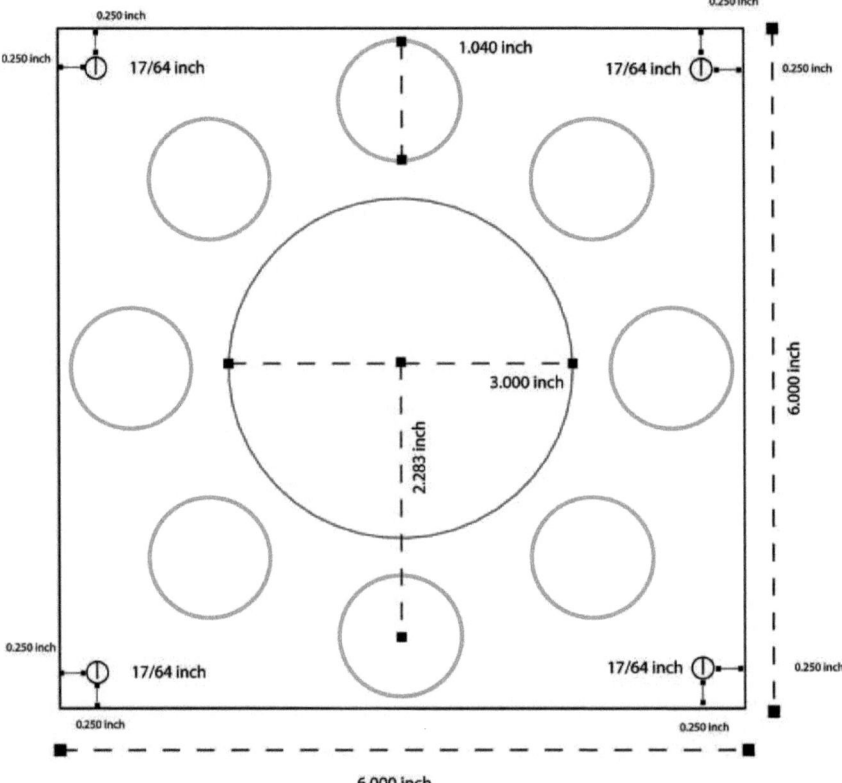

Figure 4.6: Drawing of the test detector. The holes on the edges of the HDPE matrix are for the tapped polyethylene rods, which hold together the HDPE blocks.

4.2 Neutron Detection

Figure 4.7: Different views of the test detector during the experimental phase.

4.3 Neutron Detector Construction

After reviewing the results for the test detector construction and comparing them with the computational results, the design for the neutron detector was based on the same parameter studies as for the test detector. The major differences between test and experimental detector are the number of ^{3}He proportional counters and the size of the polyethylene matrix.

The highest efficiency is achieved, when positioning two concentric rings of ^{3}He counters parallel to the beam line axis. For the so called inner ring it was quickly found that the configuration implemented for the test detector was already the optimal setting.

After finding the optimal distance for the second ring it was also determined, that one additional ring (called outer ring) with twelve ^{3}He proportional counters would be an optimal configuration. No major difference was found between twelve and fourteen proportional counters for the second ring. To keep two counters as backup counters, twelve counters were set for the outer ring.

To show that "line of sight" considerations do not play a role, the relative orientation between the inner and outer ring was varied. Different relative orientations showed only marginal differences in the overall efficiency, which one may also attribute to statistical fluctuations (compare figures 4.8 and 4.9).

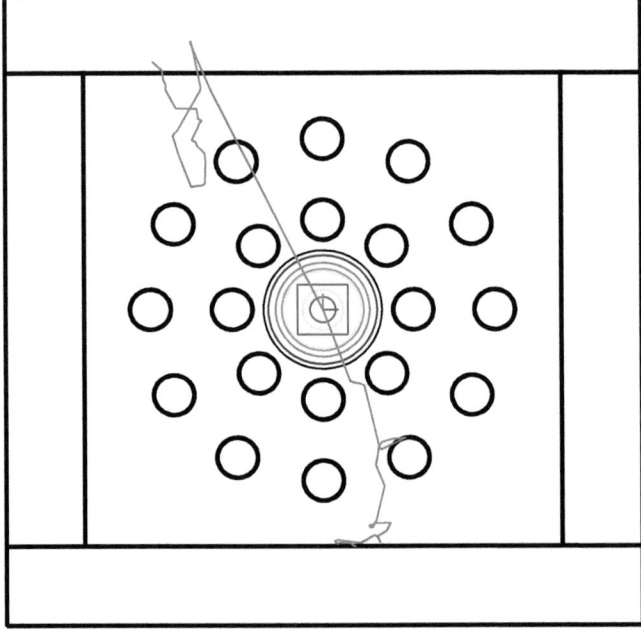

Figure 4.8: Conceptual drawing of the neutron detector with a relative ring orientation of 0 degree. The view is in beam direction. Green lines show the path of two neutrons simulated.

4.3 Neutron Detector Construction

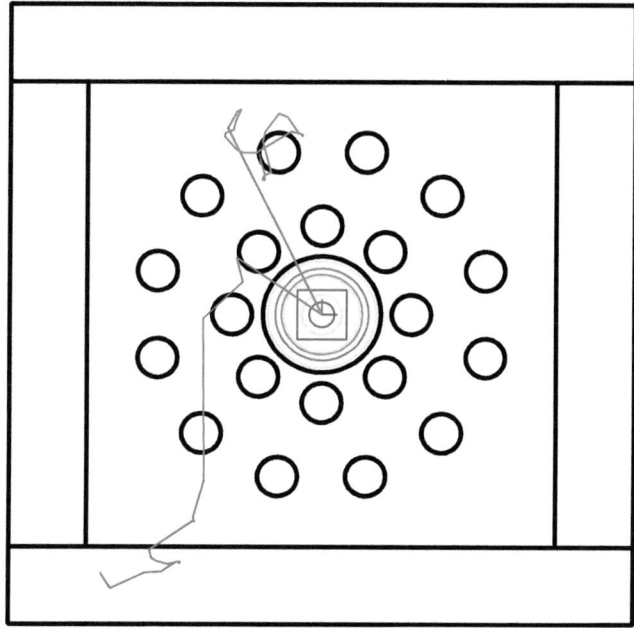

Figure 4.9: Conceptual drawing of the neutron detector with a relative ring orientation of 45 degree. The view is in beam direction. Green lines show the path of two neutrons simulated.

With the higher amount of ^{3}He proportional counters it was also necessary to increase the overall size of the polyethylene matrix. Another factor that played a role was the feasibility of the detector size, since the change of targets requires the current operator to move the detector back and forth. This led to a final size of 12 inch x 12 inch x 13 inch for the core matrix, without increasing the size to unfeasible dimensions.

After finalizing the design for the core matrix, the sensitivity toward neutrons coming from the outside into the detector was tested. Due to the higher amount of counters and the bigger amount of moderation material, the detector was intrinsically more sensitive to background radiation. Therefore, simulations were carried out, which included layers of borated polyethylene as shielding layers around the core matrix. The simulations showed, that 2 inches of borated polyethylene already deliver a sufficient background reduction.

An important feature, which is caused by a two ring design, is the fact that neutrons with identical energies are detected in the two rings with different efficiencies. Since the outer ring has a greater distance from the target than the inner ring, the neutrons therefore pass through more moderation material. Also the outer ring is automatically more sensitive toward neutrons which backscatter from the outer region of the core matrix. As a result neutrons with higher energies are more likely to be detected in the outer ring than in the inner ring. One can calculate the ratio between the detected neutrons in the inner and outer ring which provided an additional parameter for the later data analysis. For example,

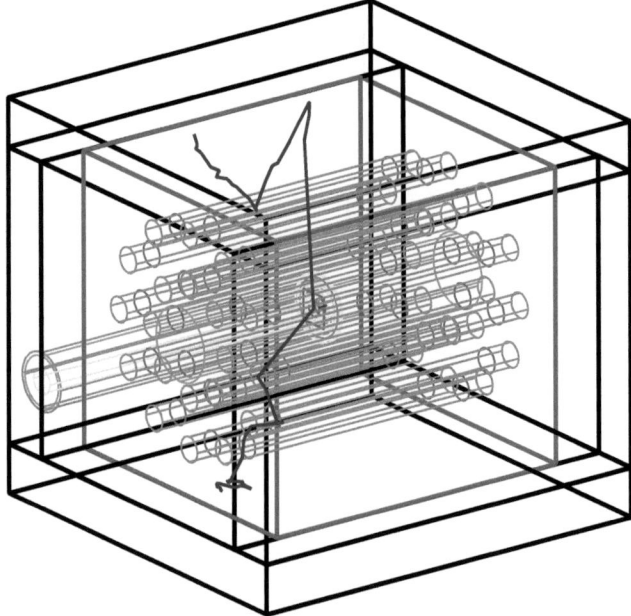

Figure 4.10: View on inner structure of the neutron detector. In blue two simulated neutron paths are shown. Black illustrates the borated Polyethylene shielding. Red is the core matrix of the detector made out of HDPE. Green and yellow are the ^{3}He counters and beampipe.

4.3 Neutron Detector Construction

if a reaction releases neutrons with higher energies than ^{25}Mg$(\alpha,n)^{28}$Si and ^{26}Mg$(\alpha,n)^{29}$Si, one should be able to observe this by calculating the ratio of of neutrons detected in the inner and outer ring (see section 5.1.3).

The detector was set up on a sliding table to allow easy target exchange and easy position measurements. To ensure that the weight of the cables and the preamplifiers does not effect the counters and dislocate them, additional polyethylene rods were inserted between preamplifier and proportional counters. Since sixty cables had to be mounted at the backside of the detector, an additional sheet of borated polyethylene was machined to hold the cables and reduce the force put onto the counters and preamplifiers.

Figure 4.11: Shown is the neutron detector during the construction phase (top) and during the experimental phase (bottom).

4.3.1 Electronics Setup of the Neutron Detectors

^{3}He proportional counters are operated at positive high voltage and deliver a signal which needs to be processed at least through two modules to deliver an interpretable signal for a data acquisition system (DAQ) (see fig. 4.12). The first module used is a preamplifier (Module A) which delivers the high voltage directly to the wire inside the counter and also delivers the signal to the next module. After that, an amplifier can be used to amplify the signal to the desired shape and height for the analog to digital converter (ADC). For the test detector this electronics setup was utilized. However, for a detector with twenty proportional counters this would have not been feasible.

Instead, the twenty proportional counters were divided into groups of four and their signals relayed to a Quad TFA (Module B). The signals were then summed to one signal (Module C). After that, the summed signal were processed through an amplifier (Module D) and relayed to an ADC (Module E). Additionally, the group sum signals were also summed up (Module F) to deliver a hardware summed signal of the inner and the outer ring. Those were then relayed to separate ADCs. Therefore it was possible to observe online the ratio between the signals coming into the five different groups and the inner and outer ring.

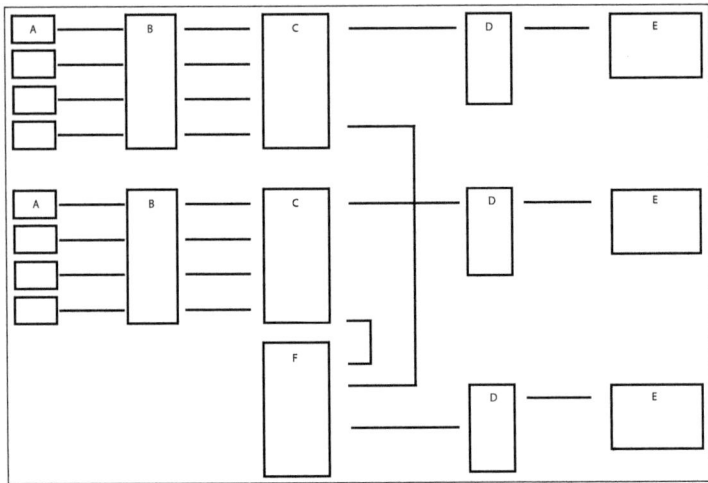

Figure 4.12: Schematic view of the neutron detector electronics for the *inner* ring. The electronics for the *outer* ring are the same with additional four ^{3}He tubes.
Module A : 2003BT Preamplifier; Module B : Ortec 863 Quad TFA; Module C : Phillips 740 Fan IN/OUT; Module D : Ortec 552 Amplifier; Module E : ADC; Module F : Phillips 740 Fan IN/OUT to sum different groups or rings.

4.3.2 Validation of Computational Calculations

An important factor in the use of computational calculations is the validation of the computational results in accordance with experimental results. To ensure a high level of reliability, it was chosen to validate the calculations in a separate manner.

4.3 Neutron Detector Construction

To be able to validate the most important parameters, such as efficiency and position sensitivity, it was chosen to make use of nuclear reactions which would serve as discrete neutron sources. This allows also to set up the complete detection system as it would be used later on in the experiment and test it towards practicality and gain experience in the operation of the set up.

Besides for accelerator calibration purposes (see section 4.1) the reaction ^{51}V(p,n)^{51}Cr has been used in studies of neutron detector performances and is relatively well known [77, 78, 79]. A useful property of the reaction is, that the produced isotope ^{51}Cr has a relatively long half-life ($t_{\frac{1}{2}}$=27.7d). Therefore, its decay to the ground-state can be conveniently measured after irradiation of a ^{51}V-target. By the detection of the prompt neutrons and the gamma activity one is able to determine the absolute efficiency of the neutron detection system via the activation method[27].

4.4 Counting Station

For decay measurements of the activated ^{51}V-targets a counting station based on a HPGe detector was built. The counting station consisted of a HPGe detector whose Ge crystal was surrounded by Pb bricks. The Pb bricks were used to reduce room background. To ensure, that between measurements, sample positioning and therefore the efficiency of the detector was not changed, a sample holder was constructed. The sample holder, mounted on the Al case of the Ge crystal, allowed to measure the efficiency of the detection system with different samples at reproducible positions.

Since ^{51}Cr decays via a single γ-decay, the efficiency at the energy of the released γ-ray (E_γ = 320 keV) is crucial. By measuring the efficiency of the detector with a calibrated ^{133}Ba source, the efficiency could be determined experimentally at 302 keV, 356 and 383 keV. Summation effects were corrected by measurements with a single γ-ray source at different distances. After the experimental determinations, the efficiency at 320 keV was interpolated to be 8%. The efficiency measurements were repeated before each experimental beam time and by different experimenters.

4.5 (p,γ)-Measurement Setup

The thickness measurements via resonances of ^{18}O, ^{25}Mg and ^{26}Mg were performed before and after the measurement of each target (see figure 4.13). The γ-rays were detected with a second germanium detector positioned at 90 degrees relative to the target chamber. The neutron detector was removed so that the germanium detector could be positioned as close as possible to the target. The crucial parameter of this setup was the reproducibility of the detector position and not the efficiency.

After the (p,γ) measurements, the Ge detector was moved as far away as possible to avoid neutron damage of the Ge crystal.

4.5 (p,γ)-Measurement Setup

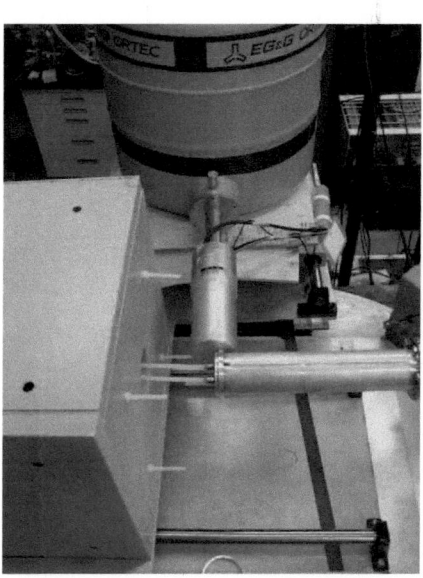

Figure 4.13: View of the (p,γ) measurement setup. For the measurement the neutron detector was moved back while the Ge detector was moved to a reproducible position. These measurements were utilized to determine target thickness and target deterioation.

4.6 Target Production

Previous measurements of the reactions ^{25}Mg$(\alpha,n)^{28}$Si and ^{26}Mg$(\alpha,n)^{29}$Si utilized primarily thermally evaporated targets, while only Wieland reports the use of implanted targets as well (see Chapter 3).

While this option was also considered, no ion beam facility could be found to produce a sufficient number of enriched magnesium targets in a financial and timely responsible fashion. Moreover, the relatively small yields and possible oxidation of the magnesium might have caused other types of problems during the experimental measurements. It was decided to use the technical possibilities at the NSL to develop reliable production procedures.

After reviewing the literature, the common technique for the production of thermally evaporated enriched magnesium targets is the so-called reduction technique[80, 81, 82, 83, 84, 85].

The technique involves the reduction of isotopically enriched (up to 99.7% in our case) magnesium-oxide (MgO) during the evaporation process with a reduction material. The advantage is that MgO is simple to purchase and with the reduction process during the evaporation one can reduce the amount of oxygen in the target. Unfortunately, literature does not give exact details on how the evaporation has to be performed. For example, evaporation temperatures, preparation of samples and backings etc. need to be considered. A major aspect of the target production is the contamination with carbon and oxygen, which effects the later measurements. As a consequence, over 100 evaporations were been performed to determine the most reliable production procedure.

For the production of ^{18}O targets thermal evaporation methods involve the evaporation of aluminium in an enriched oxygen atmosphere. The oxide Al_2O_3 is then formed and deposited on a target backing. However, due to financial and practical considerations the anodiziation method of tantalum in enriched (97%) ^{18}O water was utilized. The method was revisited and further improved by Andreas Best at the University of Notre Dame who provided the ^{18}O targets for the experimental measurements[86, 87, 88].

4.6.1 Thermal Evaporation

The term thermal evaporation describes the heating of a material in boat for our case in vacuum. The material in the boat is heated to a temperature equivalent to its vapor pressure, evaporates and then condenses on a substrate (backing).

Important for this process is that good vacuum conditions are met to avoid the reaction of the source material with residual gas. Geometry, backing material, process duration etc. have to be considered as well.

Too quick heating of the source material can lead to spontaneous evaporation which does not allow a uniform deposition on the backing material. Geometry losses of expensive target material and contamination of the vacuum chamber play an important role as well. Backing material in general has to be chosen carefully to ensure low contamination and a successful deposition of the target material.

At NSL a thermal evaporator was used, which consisted of a vacuum chamber in which the boat, the backing material and quartz-crystal monitors could be mounted. The quartz-crystal monitors were placed close to the backing and monitored the amount of deposited material. The vacuum chamber was pumped with a combined diffusion-mechanical pumping system, involving a cold trap, down to 5×10^{-7} Torr. Before each crucial evaporation

4.6 Target Production

process the holding apparatus and the vacuum chamber itself were cleaned multiple times to ensure a minimum amount of contamination. The cleaning process involved the use of chemicals such as dilute acids. Additionally, the system was pumped at least 12 hours before the evaporation was started.

4.6.2 Backing Material

For the backing material, three properties are important : purity, thermal conductivity and adhesion of the target material. For example, the evaporation of magnesium onto tantalum and copper showed different behaviors in terms of adhesion. Another aspect is that the backing material has to withstand high beam current to avoid blistering of the target itself. Following Strandberg's work, copper was chosen to be the primary backing material[89]. Due to its high thermal conductivity, Cu removes heat effectively from the target, while the target is being heated by the constant beam current.

The first tests were directed towards the adhesion of the target material onto the Cu backing, which was typically 1.5 inch x 1.5 inch x 0.04 inch. Oxygen Free High Purity Copper (OFHC) from the supplier (McMaster Carr Inc.) was used to ensure a minimum amount of contamination. Since Cu oxidizes in air, the first step was to find an optimal combination between surface cleanliness and adhesion of the target material. Strandberg et al. used first 5% aceton and later on ethanol to not only clean the surface but also etch the surface of the Cu backing. The etching of the surface by the acid will increase the adhesion of the target material onto the copper backing.

Since Strandberg et al. used natural Mg in metallic form, a series of tests with the reduction technique and ^{nat}MgO was performed. During the tests, the copper backings were first soaked in acetone and then soaked in ethanol for 20 minutes before they were transfered to the evaporation chamber. The tests showed that only about 70% of all evaporations resulted in a successful adhesion of the target material to the Cu.

To clean the surface of a Cu part, Aalseth and others have reported the treatment of the surface with a activation-passivation method[90, 91]. The method utilizes H_2O_2 to remove oxides and Cu^+/Cu^{2+} residuals (Fenton mechanism) :

$$Cu^+ + H_2O_2 \rightarrow Cu^{2+} + OH^- + OH^{\cdot} \tag{4.13}$$

$$Cu^{2+} + H_2O_2 \rightarrow Cu^+ + H^+ + OOH^{\cdot} \tag{4.14}$$

$$OH^{\cdot} + H_2O_2 \rightarrow H_2O + OOH^{\cdot} \tag{4.15}$$

$$OH^{\cdot} + Cu^+ \rightarrow Cu^{2+} + OH^- \tag{4.16}$$

The produced hydroxyl radical acts as the cleaning species and the Cu ions remain in solution. To prevent a re-oxidation after the treatment with H_2O_2, the sample was put into citric acid solution, which is known as passivation. The sample was held into both solutions for about 20 minutes each and showed already optically impressive results. In order to increase the purity, the samples were then baked out over several hours in a vacuum oven to evaporate residual humidity and contaminants.

This procedure showed an improvement in terms of adhesion compared to the method used by Strandberg et al. since almost all evaporation processes were successful. The use of a vacuum oven did not result in noticeable improvements.

The first beam tests showed large levels of carbon contaminaton, probably due to carbon inclusions deeper in the backing. Since the ion beam penetrates through the target material and is stopped in the copper backing, those type of impurities can not be avoided.

An improvement in terms of impurities is the use of gold-plated copper disks[45, 92]. OFHC disks are put into an electroplating bath and are then coated with an Au layer. To avoid diffusion of the gold into the copper, a $1\mu m$ thick nickel layer was plated onto the copper first. To stop the α-beam in the Au layer and not in the Ni or Cu layer, a sufficiently thick layer of Au is necessary. Calculations were performed in which the implantation of α-particles into Au was simulated. For this purpose the software SRIM2008 was used ([93], also see http://www.srim.org).

The calculations showed that for a maximum beam energy of $E_\alpha = 3$ MeV a $5\mu m$ layer of Au would be sufficient to stop the beam. The OFHC disks were obtained by Goodfellow Inc. corporation while the electrochemical plating was performed at the Max-Planck-Institut für Chemie (MPIC).

The Au plated copper disks were also treated by the activation-passivation method, for several minutes, and then directly transferred to the evaporation chamber. In terms of adhesion they showed a perfect behaviour and turned out to be the backings with the least contamination.

4.6.3 Evaporation Process

In literature, the evaporation process of MgO being reduced to Mg includes a mixture of MgO with a material that reduces MgO during the process.

Tests with tantalum, zirconium and titanium showed, that Ta is the preferred reduction material, since it showed no difficulties in handling and processing. To reduce costs and contamination of the evaporation chamber, the amount needed for the evaporation process was first determined by weighing the MgO and the Ta before mixing it.

The amount of tantalum used is critical for the reduction process. The amount of oxygen deposited on the target backings could be reduced by simply increasing the amount of Tantalum compared to a fixed amount of MgO. This was shown with the RBS measurements as well. The reduction process can be described as :

$$10 \text{ MgO} + 4 \text{ Ta} \rightarrow 10 \text{ Mg} + 2 \text{ Ta}_2\text{O}_5 \tag{4.17}$$

and therefore it can be concluded that, with an increased amount of reduction partners, the reduction process can be more effective. It was found to be optimal to have an absolute mass ratio of 25:1 for Ta:MgO. MgO amounts of 15 mg were sufficient for one target production process.

Several tests with different heating sources (boats) were also conducted. Boats come in different materials and shapes. Boats made out of tantalum can only reach temperatures up to 1600 °C while boats made out of tungsten could reach temperatures up to 1800 °C. This is critical since no exact evaporation temperature (or vapor pressure) has been reported for the reduction process and boat breakdowns were observed during the first test. Therefore, the first series of tests conducted included the use of so called dimple boats made out of tantalum, tungsten and molybdenum (see figures 4.14 and 4.15).

Surprisingly, the molybdenum boats showed the best results in achieving an evaporation. This was possibly also enhanced by the molybdenum itself, since it can reduce MgO by itself as well[94].

A different aspect is the surface contact between the MgO, Ta and the boat. It was first increased by using a few drops of water and mix the MgO and Ta in the boat itself. Unfortunately, during the heating process the water evaporates suddenly and the mixture

4.6 Target Production

Figure 4.14: Sketch of a dimple boat (from Lesker Inc.).

explodes out of the boat. Eventually ethanol was used, which evaporates at room temperature. Still, the heating had to be performed very slowly at the beginning to avoid any further explosions.
This effect could be monitored via an ionization gauge which measured the pressure inside the chamber. For all evaporations the pressure initially rose and then dropped again, indicating the evaporation of the alcohol. Until the evaporation of the magnesium set in, the pressure fell back and stayed stable. Additionally, a shutter was utilized to avoid contamination of the backings during the pumping and heating process. Once the evaporation set in, the shutter was removed.

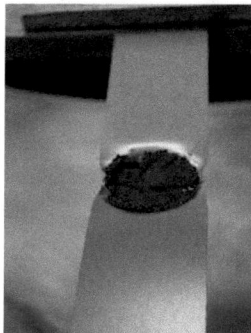

Figure 4.15: Picture of the MgOTa mixture in a dimple boat before evacuating the evaporation chamber.

For the production of RBS targets a carbon disk was placed as close to the Au backing as possible, while the boat remained 11 cm below the backings to ensure a deposition on both backings. The RBS measurements of the targets produced showed an increased amount of tantalum, molybdenum and oxygen. An explanation is, that during the reduction process Mg breaks out of the mixture and MoO as well as Ta_2O_5 get deposited onto the backing as well. While Ta_2O_5 is the product of the reduction process, MoO can be produced as reaction product between the Mo boat and MgO.
For a tighter geometry and better surface contact between boat and target mixture, differently shaped boats were tested. So called pinhole boats showed an improvement compared to the dimple boats (see fig. 4.16). The narrow opening of the pinhole itself leads to a too narrow distribution of the target material and makes it difficult to influence the mixture of the target material in the boat, however.

Figure 4.16: Sketch of a pinhole boat (from Lesker Inc.).

A better solution was found in the use of so called close end tube heaters provided by Lesker Inc. (fig. 4.17). The mixture can be easily put into these type of boats and the geometry of the boat allows a tight distribution of the target mixture. Before each evaporation process the tube heaters were cleaned and then baked at very high temperatures.

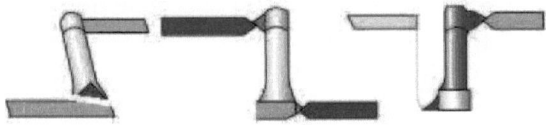

Figure 4.17: Sketch of a tube heater (from Lesker Inc.).

The targets made with tube heaters showed the best results in terms of contamination from other elements and were used for the production of all magnesium targets (see figures 4.18 and 4.19).

After the evaporation process, the vacuum chamber was flooded with argon to minimize oxidation through the venting process. Additionally, both, the production and the RBS target, were directly placed on target holders and transported to the experimental setup under an Ar atmosphere.

4.6 Target Production

Figure 4.18: Picture of a Mg target after evaporation.

4.6.4 Vanadium Targets

For the activation measurements a series of vanadium targets were produced. The evaporation of vanadium compared to magnesium oxide is relatively simple. The vanadium can be placed in a tungsten dimple boat and then heated until it evaporates onto a previously baked Ta backing. Since vanadium has a high vapor pressure, it is important to have good surface contact between vanadium and the boat to ensure that the heat is well-distributed.

Figure 4.19: Picture of the evaporation setup within the evaporation chamber. Shown are the AuNiCu and RBS backing mounted next to each other. The shutter is in the "closed" position to avoid contamination, while the tube heater is mounted on the current leads.

Chapter 5

Experimental Results

5.1 Neutron Detector Performance

5.1.1 Positioning

Before each experimental beam time the optimum position of the detector relative to the target was determined. This was achieved by placing the detector on a sliding table and moving it along the beam line axis. For each position, the neutron yield (number of recorded neutrons divided by incident particles) was recorded. Each position was measured several times to ensure reproducibility.

In figure 5.1 the recorded neutron yield at each test detector position is shown. As expected, at the center of the detector block one achieves the highest yield. With less space covered by the neutron detector relative to the target, the yield is decreased. As shown in a range of about 4 inch around the center position the detector performance is not influenced by possible misalignments.

Additionally, the results from the computational simulations were normalized to the data at the detector center position. For the inner region of the detector the computational simulations are in accordance with the experimental measurements. For positions where the target is in the outer regions of the detector, however, there are differences between the simulations and experiments. A possible reason is the treatment of the neutron moderation at the transition between air and polyethylene. However, the obtained results were found to be sufficient for the experimental handling of the neutron detector.

5.1.2 Efficiency

The efficiency of the neutron detector was determined by activation measurements using the reaction ^{51}V(p,n)^{51}Cr. Freshly produced ^{51}V targets were irradiated with a fixed proton energy E_p. This led to isotropically, monoenergetic neutrons emerging from the center of the neutron detector. The efficiency can then be determined by measuring the produced activity in the previously described counting station.

The efficiency is defined as the ratio of the number of neutrons detected (n_d, which is the sum of the neutrons detected in inner ring and outer ring) to the number of released neutrons (n_r) :

$$\epsilon = \frac{n_d}{n_r} \qquad (5.1)$$

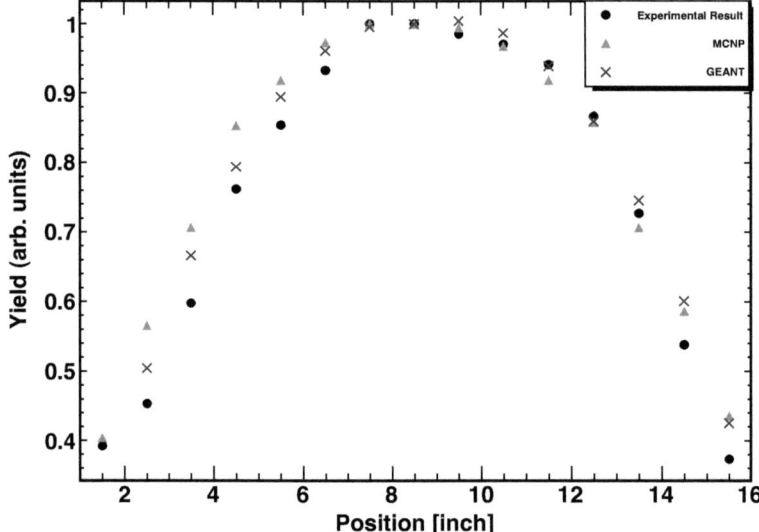

Figure 5.1: Shown is the recorded neutron yield versus the detector position relative to the target. Position 8 marks the target being in the center of the detector. Distances between adjacent positions are one inch. The different data sets are normalized to the one at the center position.

5.1 Neutron Detector Performance

The number of neutrons detected was measured promptly and was corrected for background radiation and dead time effects, while the number of released neutrons is equal to the number of produced ^{51}Cr nuclei. The number of ^{51}Cr nuclei produced and therefore the number of released neutrons was determined in turn, by measuring the activity of the activated ^{51}V target. The number of ^{51}Cr produced in a single activation (N_i); and (still) present a some t is given by :

$$N_i = \frac{k_i \cdot P}{\lambda}(1 - e^{-\lambda \cdot t_s}) \cdot e^{-\lambda \cdot (t-t_e)} \qquad (5.2)$$

Here λ is the decay constant and P the production rate. t_s represents the start time of the activation and t_e the end time. The factor k_i is a correction factor for possible target depletion and beam current deviations. It was determined by calculating the neutron yield after each activation period. Each ^{51}V target was irradiated multiple times, to be able to monitor the data online and to be able to react e.g. to target deterioation. The index i denotes multiple activation intervals for one ^{51}V target.

The production rate P was determined via the measurement of the produced ^{51}Cr nuclei. The number of emitted γ-rays (N_γ) from the ^{51}Cr nuclei during a defined time interval is given by :

$$N_\gamma = \frac{N_c}{I \cdot \epsilon_{Ge} \cdot DT} \qquad (5.3)$$

with N_c being the number of recorded γ-rays, I the branching ratio, ϵ_{Ge} the efficiency of the Germanium detector and DT the dead time correction.

N_γ is simply the difference of the number of ^{51}Cr nuclei at the beginning ($N_i(t_{AS})$) and the end ($N_i(t_{AE})$) of the measurement. This leads then to :

$$\sum_i (N_i(t_{AS}) - N_i(t_{AE})) = \frac{N_c}{I \cdot \epsilon_{Ge} \cdot DT} \qquad (5.4)$$

Inserting equation 5.2 into equation 5.4 results in :

$$\frac{N_c}{I \cdot \epsilon_{Ge} \cdot DT} = \sum_i \left(\frac{k_i \cdot P}{\lambda}(1 - e^{-\lambda \cdot t_s}) \cdot e^{-\lambda \cdot (t_{AS}-t_e)} - \frac{k_i \cdot P}{\lambda}(1 - e^{-\lambda \cdot t_s}) \cdot e^{-\lambda \cdot (t_{AE}-t_e)} \right) \qquad (5.5)$$

By assuming that the production rate is constant for each activation interval, equation 5.5 can be transformed to :

$$P = \frac{N_c}{I \cdot \epsilon_{Ge} \cdot DT} \cdot \frac{\lambda}{\sum_i k_i \cdot (1 - e^{-\lambda \cdot t_s}) \cdot (e^{-\lambda \cdot (t_{AS}-t_e)} - e^{-\lambda \cdot (t_{AE}-t_e)})} \qquad (5.6)$$

Multiplying the production rate with the activation time t_A results in the number of neutrons released over each activation interval. Hence, the efficiency can be expressed as :

$$\epsilon = \frac{n_d}{P \cdot t_A} \qquad (5.7)$$

By inserting equation 5.6 into equation 5.7 one then calculates the efficiency for a single neutron energy.

The determination of the efficiency at neutron energies higher than 650 keV was not possible due to very high dead time effects in the detection system.

To calculate the efficiency at higher energies, computational simulations were used. While the simulations were scaled to match the experimental results up to 650 keV, trends of the simulations and the experimental results do agree (see figures 5.2 and 5.3).

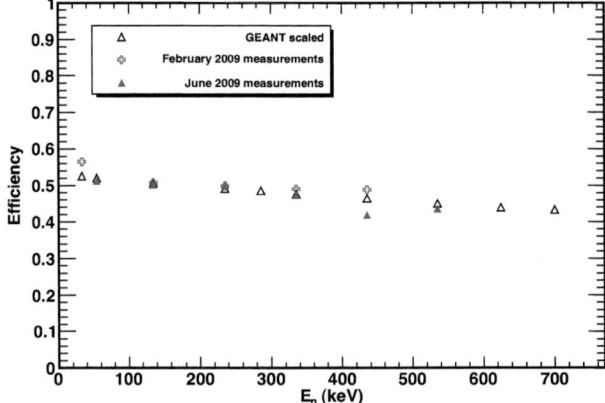

Figure 5.2: Shown are the measured efficiency data of the neutron detector in comparison to the scaled efficiency data obtained with GEANT4.

Since the neutron energies during the experiments were not limited to the energy range below 650 keV, the efficiency for neutron energies up to 10 MeV was derived by fitting a polynomial efficiency function.
Surprisingly, using a single efficiency function it was not possible to reproduce satisfactorily the simulated data. As shown in figures 5.4 and 5.5, at energies above 3500 keV the behaviour of the efficiency is different from that of energies below 3500 keV. This unexpected behaviour has not been reported before.

Effect of (n,n) Resonances on the Detector Performance

Surprisingly, when increasing the number of energy steps within our computational simulations (finer energy resolution), structures within the efficiency curve became evident. This behaviour was unexpected, since e.g. the cross section of the ^{3}He(n,p)^{3}H reaction shows a flat, non-resonant behaviour (see e.g figure 4.2). Only Evans reported a similiar behaviour, but for a lower energy region and was not able to make a convincing statement[95].
Several tests were performed to investigate the effect. Statistical fluctuation, the direction of the neutrons and other factors were excluded by varying those parameters and observing no effect on the results.
After reviewing experimental data on elastic (n,n) scattering, it became evident that most likely ^{nat}C(n,n) resonances cause the peak structures within the simulated data. This could be confirmed by setting the cross section data used by the simulation frameworks to zero. In comparison to the simulations with the full set of cross section data, the

5.1 Neutron Detector Performance

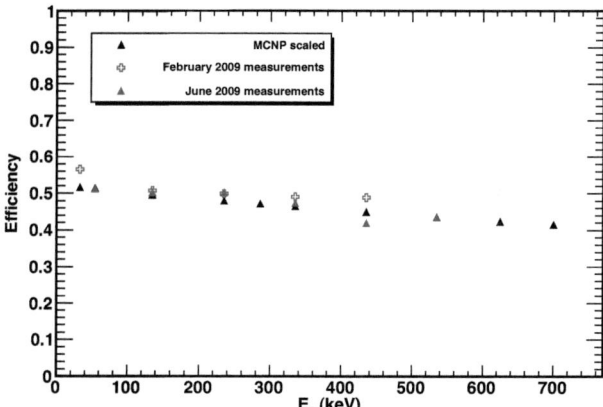

Figure 5.3: Shown are the measured efficiency data of the neutron detector in comparison to the scaled efficiency data obtained with MCNP.

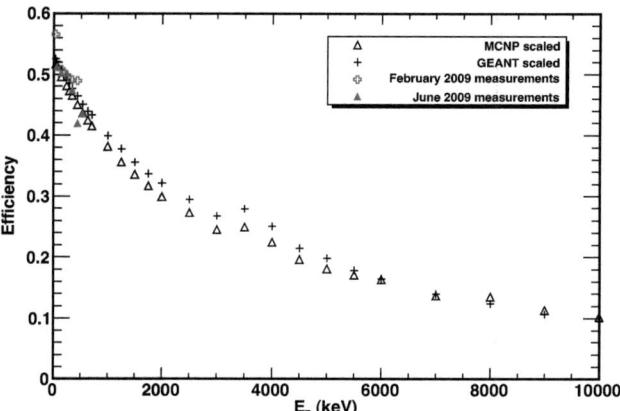

Figure 5.4: Scaled results obtained with computational simulations and the experimentally obtained neutron detector efficiency as a function of initial neutron energy up to 10 MeV.

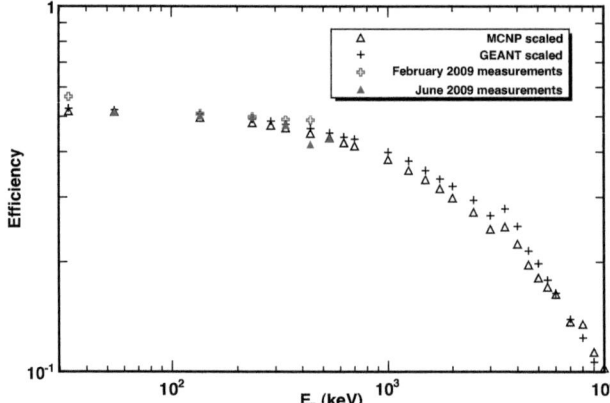

Figure 5.5: Experimentally obtained neutron detector efficiency and scaled results from computational simulations as a function of initial neutron energy up to 10 MeV on a log-log scale.

simulations with $\sigma_{el}(^{nat}C(n,n))=0$ showed a flat efficiency behaviour (see figure 5.6).
It is important to note that not the inelastic scattering cross section is responsible for the occurrence of these resonance effects.
Instead, as a neutron enters the polyethylene matrix, it experiences elastic (n,n) scattering as well. As a results, the path length of the neutron within the moderator material is extended and the probability of the neutron experiencing inelastic scattering (and therefore moderation) is enhanced. This influences directly the computational simulations (see section 4.2.3).
Moreover, one has also to take into account the fact that a neutron entering the moderator material with an energy higher than a specific (n,n) resonance eventually will experience an elastic scattering resonance due to the previous moderation.
While beyond the scope of this thesis, it should be noted that an experimental validation for the effects described above is possible. This can be realized, e.g., using the the reaction $^{7}Li(p,n)^{7}Be$ for which the cross section is constant within a few percent for certain energy regions (e.g. [96]). Measuring the neutron yield over such an energy region, the neutron detector response should show the previously described resonance effects. In any case, the neutrons from the (α,n) reactions of interest are not affected and the effect does play a major role in the further considerations.

5.1 Neutron Detector Performance

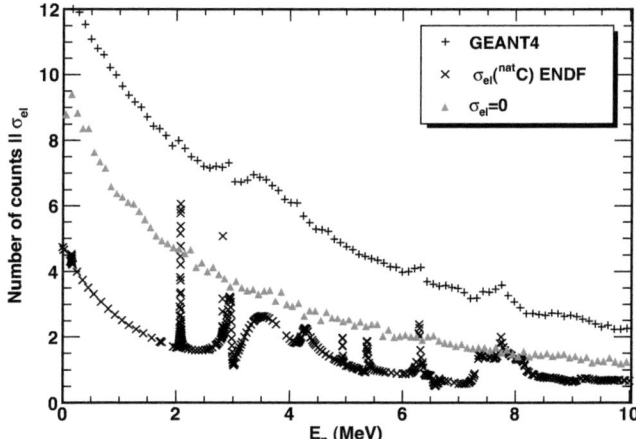

Figure 5.6: Illustrated are the results from the GEANT4 simulation for the detection efficiency, the elastic (n,n) cross section data and the GEANT4 simulation results with $\sigma_{el}(^{nat}\text{C(n,n)})=0$.

Efficiency Function

Following the match of the simulation data with the experimental data, an efficiency function was derived by fitting the data with two polynomial functions. For $E_n \leq 3$ MeV the efficiency is described by :

$$\epsilon = 0.52311 - 1.928 \cdot 10^{-4} \cdot E_n + 5.856 \cdot 10^{-8} \cdot E_n^2 - 5.574 \cdot 10^{-12} \cdot E_n^3 - 3.2358 \cdot 10^{-15} \cdot E_n^4 \tag{5.8}$$

while the efficiency function for $E_n \geq 3$ MeV is :

$$\epsilon = 0.6081 - 1.58 \cdot 10^{-4} \cdot E_n + 1.847 \cdot 10^{-8} \cdot E_n^2 - 7.745 \cdot 10^{-13} \cdot E_n^3 \tag{5.9}$$

where the neutron energy E_n is given in keV. The efficiency functions derived from the MCNP5 or GEANT4 data turned out to be equal within statistical uncertainties.

Comparison to Previous Detection Systems

As mentioned before, similar detection systems have been used in the past. In figure 5.7 a comparison between different detector setups is shown. As one can observe, for lower neutron energies the detection system built for this thesis observes has a higher efficiency.

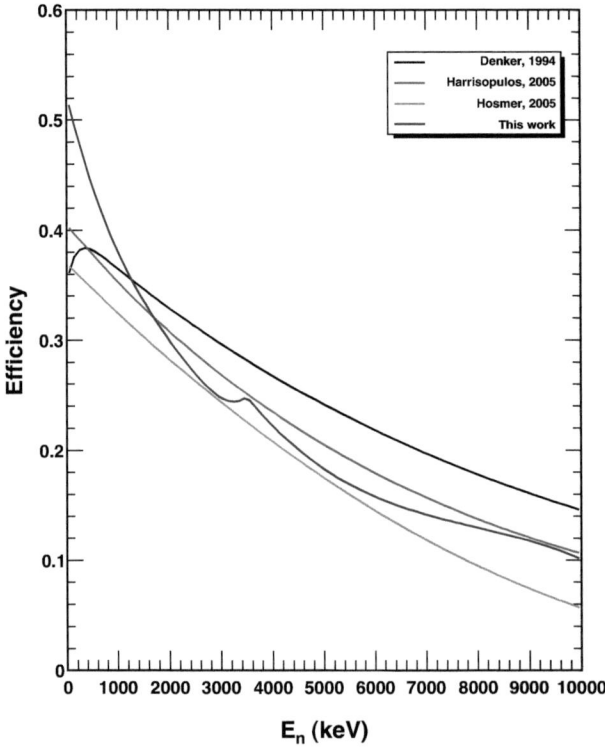

Figure 5.7: Shown are the different efficiency functions of similar detections systems compared to the system developed here[8, 97, 98].

5.1.3 Ring Ratio

The ring ratio (R) is defined as the ratio between the number of counts detected within the inner ring (N_{IR}) and the number of counts within the outer ring (N_{OR}):

$$R = \frac{N_{IR}}{N_{OR}} \tag{5.10}$$

The number of detected neutrons within each ring is dependent on the initial neutron energy. The higher the initial neutron energy, the longer the path through the moderator material. Consequently, the ring ratio can be used as a tool to roughly estimate the initial neutron energy.
The ring ratio was calculated from the efficiency measurements and compared to the results of the computational simulations (see figure 5.8). The efficiency for each neutron energy can then be assigned to a specific initial neutron energy as well (see figure 5.10).

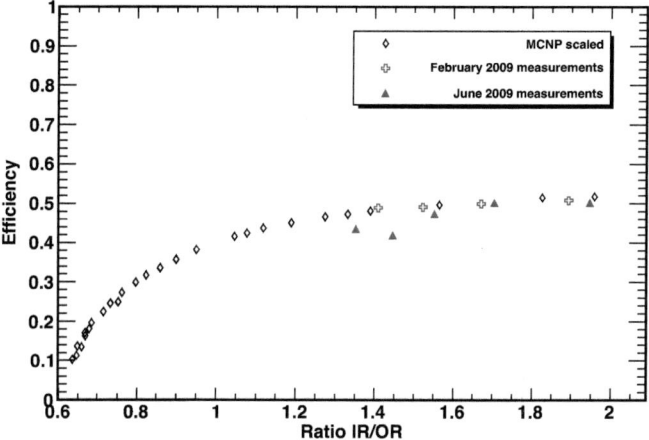

Figure 5.8: Plotted is the effciency of the neutron detector as a function of the ring ratio. The results obtained with MCNP provide the best match to the experimental data.

The nature of (α,n) reactions as two step processes involves also the occurrence of so-called neutron groups. A neutron group n_i describes the neutrons who populate the ith state of the final nucleus. Neutrons belonging to different groups have different energies (see figure 5.9). This results in a neutron detection efficiency that is different for each neutron group. This can be corrected for by analyzing the ring ratios and determining the neutron groups. A function for the neutron energy was derived by fitting the data displayed in figures 5.10 and 5.11 :

$$E_n(R) = e^{(11.42-3.762*R)} \tag{5.11}$$

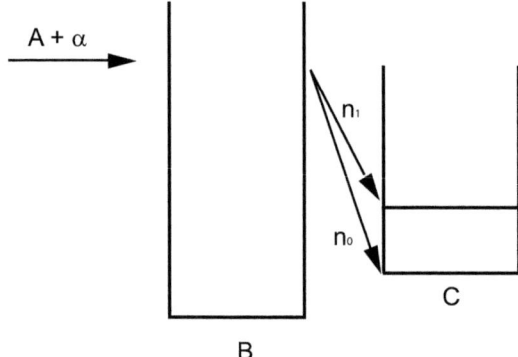

Figure 5.9: Schematic drawing of a compound reaction, illustrating the occurence of different neutron groups.

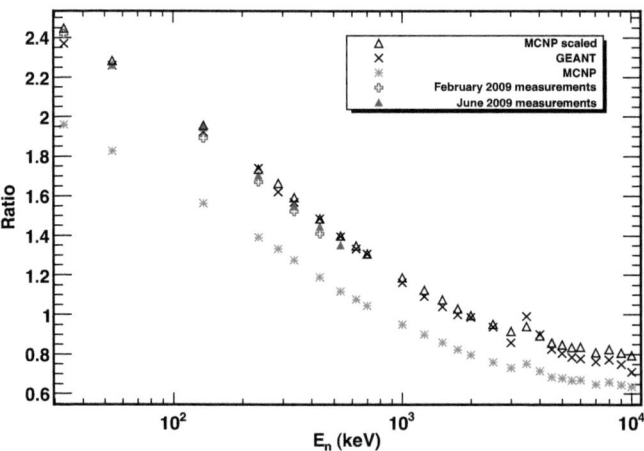

Figure 5.10: Plotted is the ring ratio versus the initial neutron energy. Note, that MCNP does not reproduce the ring ratio as well as GEANT. The MCNP results wered adjusted with a scaling factor therefore. The data also shows indication for a peak at energies above 3500 keV. For details see section 5.1.2.

5.1 Neutron Detector Performance

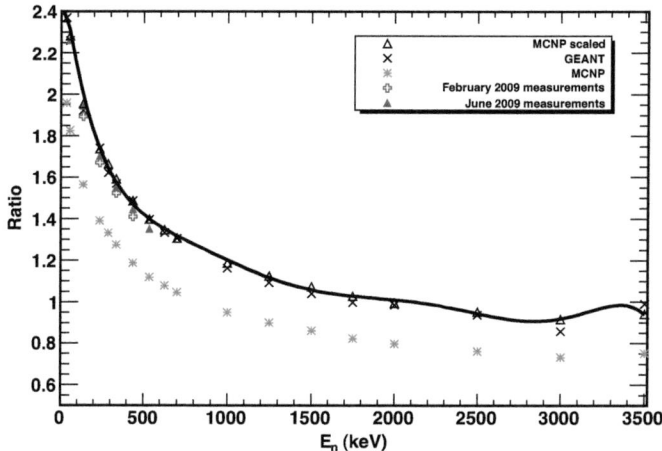

Figure 5.11: Plotted is the ring ratio versus the initial neutron energy up to an initial neutron energy of 3500 keV. The solid line represents the fit function derived.

5.2 RBS Measurements

After each Mg target production process, two targets were available. One target consisted of the enriched Mg layer on a gold backing (production target) while one target consisted of a carbon disk as a backing (RBS target). Both targets were mounted at their target stations immediately after being produced. The RBS targets were measured first to determine the contamination level. The RBS measurements then indicate the level of contamination by oxygen and other elements heavier than carbon. This allowed to determine the effects on the production using different equipment and procedures. Additionally, the used stopping powers for the data analysis of the production data could be adjusted accordingly.

An important results was that an increased amount of Ta within the MgO-Ta mixture reduces the oxygen level. Ocurrence of Mo within the target layer was observed as well, when Mo boats were used.

The RBS measurements showed that a virtually oxygen-free production of Mg target layers is in fact possible. The main problem appears to be the process point where targets are transferred to their respective target station and are exposed to air. Re-oxidation of the evaporated Mg layers already during the evaporation process seems to be possible as well, since Mg already oxidizes at vacuum levels below 10^{-7} Pa[99]. A significant oxidation during the measurement of a production target was not observed.

5.2 RBS Measurements

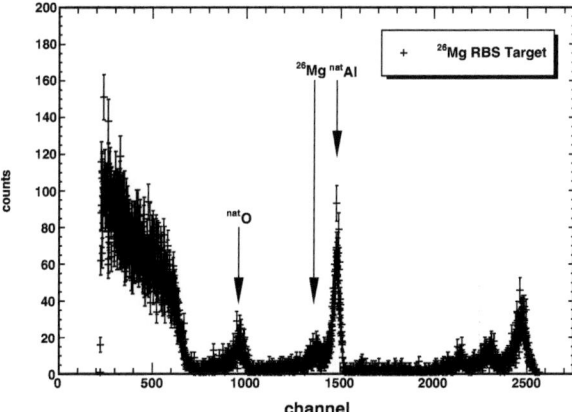

Figure 5.12: RBS spectrum of a contaminated ^{26}Mg target. The peaks at channel numbers higher than 2000 can be assigned to high mass impurities such as Mo or Ta. The Al peak is probably caused by the sample holder in the target chamber.

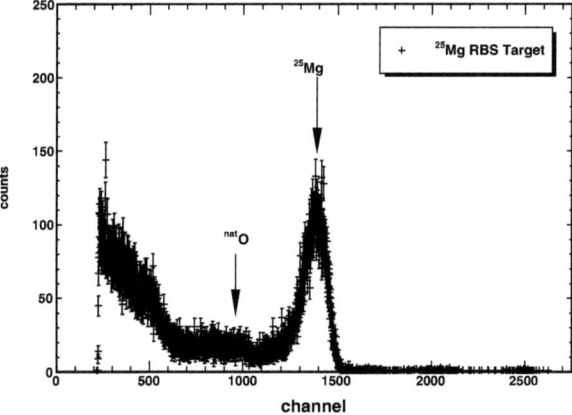

Figure 5.13: RBS spectrum of a relatively clean ^{25}Mg target. Compared to the target in figure 5.12 the reduced oxygen content and absence of high mass impurities is evident.

5.3 Target Thickness Measurements

The target thickness of each magnesium production target was determined via a (p,γ) resonance measurement. The full width maximum of the resulting peak in the excitation curve displays the target thickness[27]. The thickness of the magnesium targets varied, while the target thickness of the oxygen targets was kept the same (see tables 5.1 and 5.2). This was due to the high reproducibility of the anodizing process as compared to the fluctuations within the evaporation process by which the Mg targets were produced.

Target Isotope	E_r	E_γ^1	E_γ^2
^{24}Mg	823	2611	3061
^{25}Mg	775	417	3891
^{26}Mg	718	840	1016

Table 5.1: List of resonances used for the target thickness measurements on the Mg production targets and their respective key γ-ray transitions.

Target	Δ (2 MeV) [keV]	Δ [μg/cm^2]	Mg : O [atom/atom]
^{24}Mg 2	28.86	26.21	1.18:1
^{25}Mg 1	23.11	20.94	3.09:1
^{25}Mg 2	41.98	38.57	2.74:1
^{26}Mg 1	14.61	14.31	1.35:1
^{26}Mg 2	31.63	30.98	2.30:1
^{18}O 1-6	5.80	11.5	—

Table 5.2: Target thickness and Mg:O ratios for the production targets.

5.4 Background Correction Techniques

An intrinsic feature of detection systems based on neutron moderation and counting is the loss of energy information. As the neutrons are detected, one is not able to assign each neutron to a specific reaction. Possible contamination of the targets leads to background reactions influencing the experimental data. Especially if one compares the cross sections of the background reactions and reactions of interest it becomes evident that often even smallest impurities can lead to problems during the data analysis (see figure 5.14). Here, with the Mg targets already contaminated with oxygen and the occurrence of carbon in almost any material, there is no doubt of the occurence of background reactions such as $^{13}C(\alpha,n)^{16}O$, $^{17}O(\alpha,n)^{20}Ne$ and $^{18}O(\alpha,n)^{21}Ne$.

It has to be noted that not only contamination within the targets but also e.g. on the beamline slits can lead to an increase in background reactions. This was illustrated by simply increasing the beam current on the beamline slits and recording the neutron yield at a fixed energy. As a consequence, the current on the beamline slits was kept below 0.5 μA.

As shown in figure 5.15, the raw experimental data already allowed to identify most resonance peaks and to assign them to specific background reactions. Two techniques were employed to correct the experimental data.

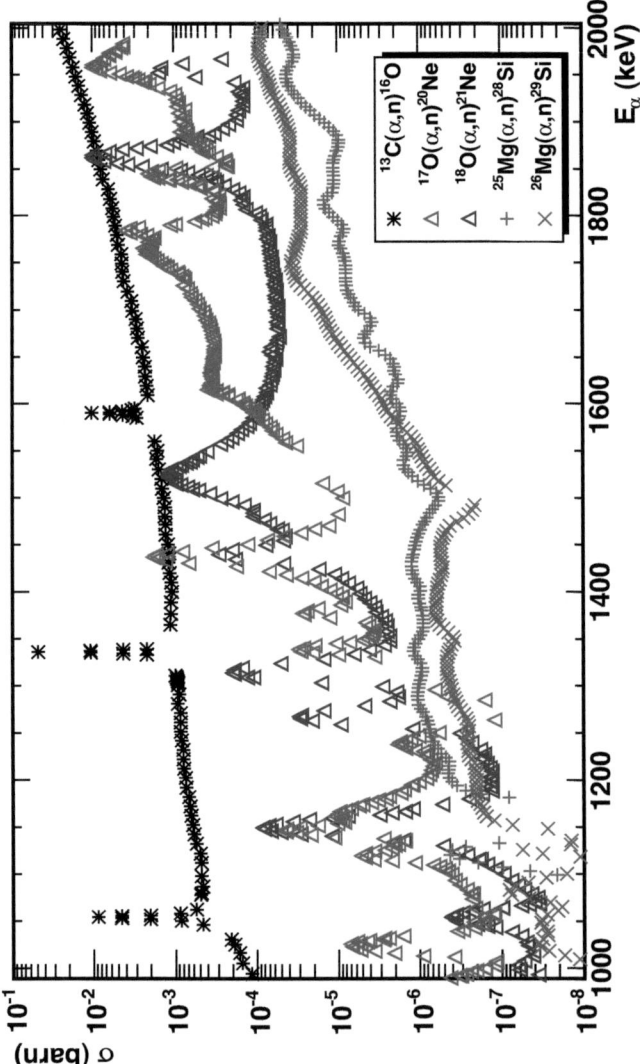

Figure 5.14: Comparison of the cross sections for relevant reactions ion this work including background reactions[57],[98].

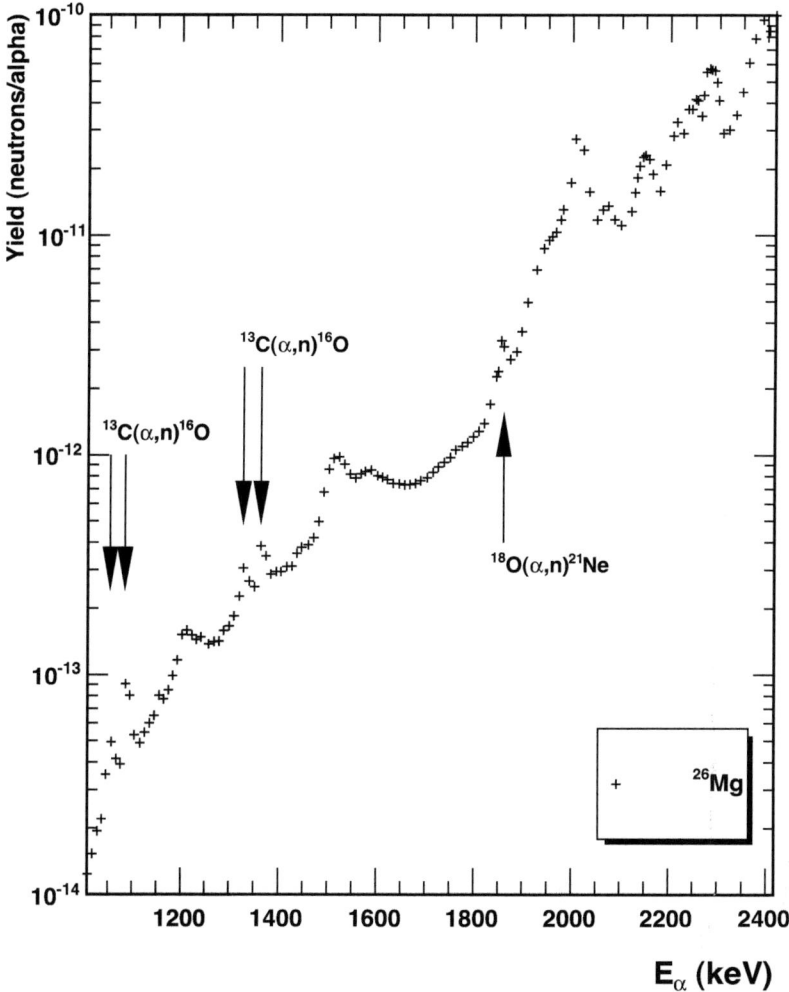

Figure 5.15: Yield curve for ^{26}Mg$(\alpha,n)^{29}$Si obtained with an enriched ^{26}Mg target. The arrows show the positions of resonance peaks caused by the background reactions ^{13}C$(\alpha,n)^{16}$O and ^{18}O$(\alpha,n)^{21}$Ne.

5.4.1 Background Targets

The first technique involved the production of *background* targets. A background target was produced in the same way as a production target, but consisting of "impurities" only. For the Mg production targets an isotopically enriched (99.75 %) ^{24}Mg target served as an background target. This was possible since the ^{24}Mg(α,n)^{27}Si reaction occurs only at energies outside the experimental energy range (E_{thres} = 8396 keV). For the ^{18}O production targets a baked Ta backing was used as a background target.

^{18}O background target

The measurements for the ^{18}O background target showed only the ^{13}C(α,n)^{16}O reaction as an additional contribution to the experimental data (see figures 5.16 and 5.17). The ^{13}C(α,n)^{16}O reaction, in general, was identified via the ^{13}C(α,n)^{16}O resonance peak at E_r = 1053 keV.

A ROOT analysis code was developed executing the following steps :

- Identify peak position of E_r = 1053 keV of ^{13}C(α,n)^{16}O reaction.

  The peak positions of the ^{13}C(α,n)^{16}O reaction and the production target varied for different targets. This is mainly due to the fact, that the contamination may be situated at different geometrical locations.

- Scale peak height of E_α = 1053 keV peak

  Thickness and the level of the contamination directly influence the peak height in the experimental data. Therefore, a scaling routine was implemented to match the contamination levels.

- Match energy of background and production data

  The energies at which background and production data are measured do not match exactly. A routine was implemented to evaluate the background data at any given energy based on the experimental background data.

- Subtract evaluated background data from experimental data

  Once background yield and energy binning are matched, the evaluated yield was subtracted from the production data.

This routine was used for the directly recorded neutron yield Y_r (which is already corrected for dead time effects and cosmic background radiation).

In figure 5.18 an example is shown, in which the uncorrected data, the evaluated background target data and the corrected data are shown. It is evident that each target shows a noticeable, but different level of ^{13}C contamination, despite identical target production process and handling. The ^{13}C contamination only had a noticeable effect on the experimental data at energies below E_α = 1400 keV. The peak width of the ^{13}C(α,n)^{16}O resonance peak also indicates a relatively thin target contamination.

5.4 Background Correction Techniques

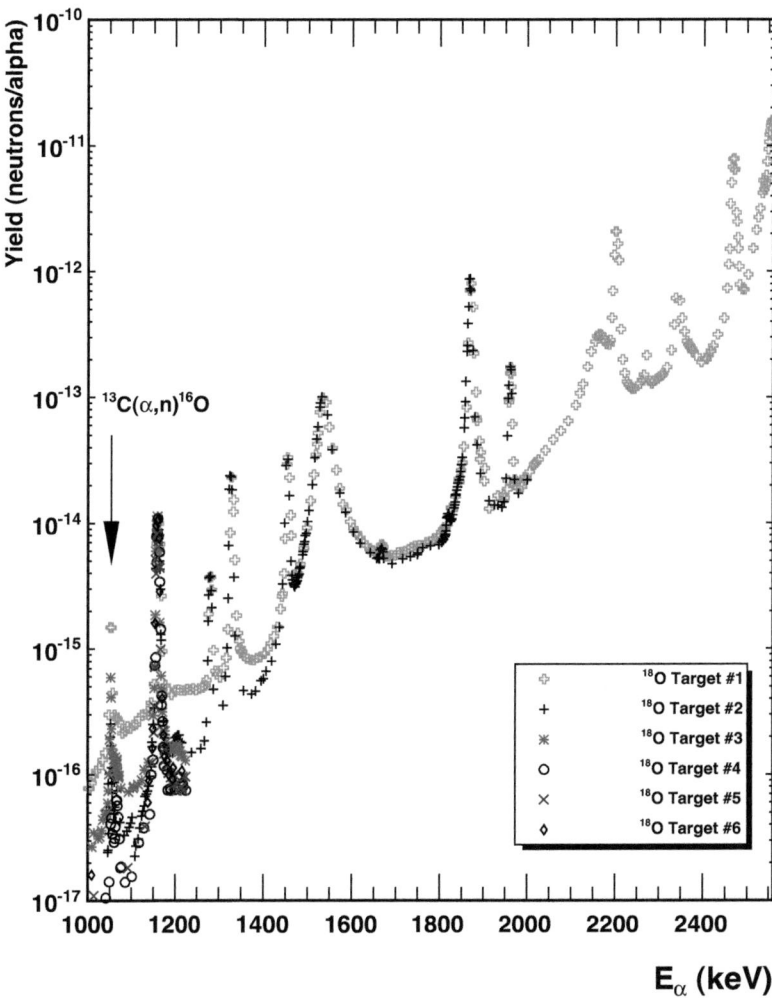

Figure 5.16: Yield curves for $^{18}O(\alpha,n)^{21}Ne$ obtained with enriched ^{18}O targets. The arrow shows the position of the $E_r = 1053$ keV resonance peak of the reaction $^{13}C(\alpha,n)^{16}O$.

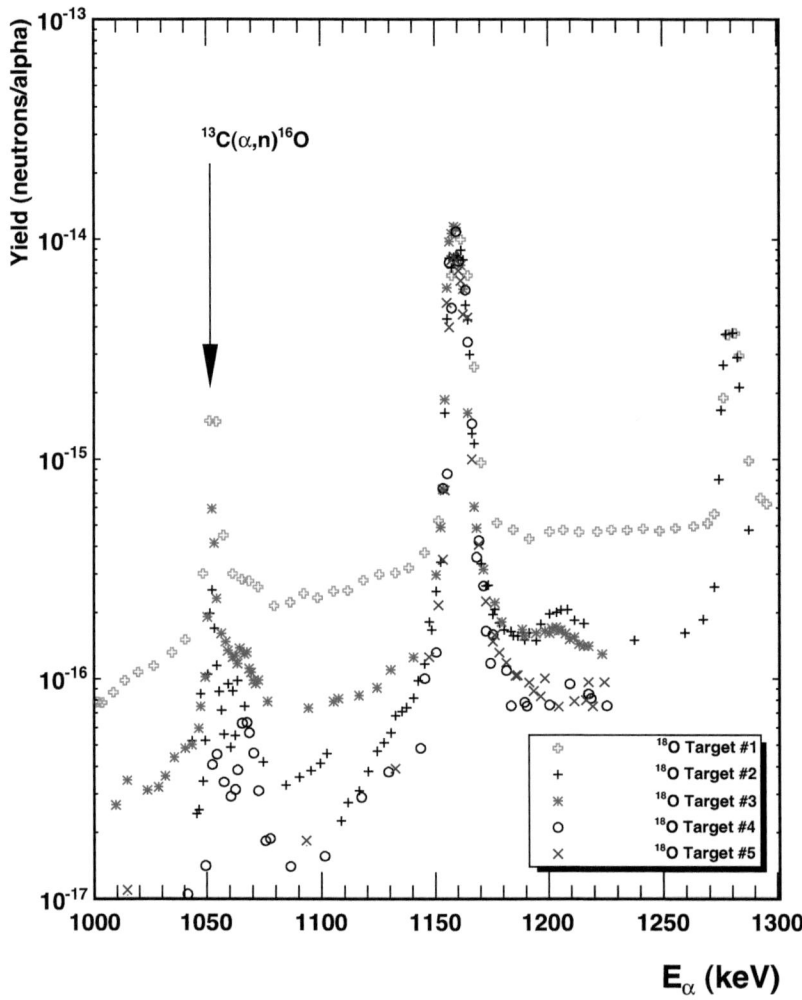

Figure 5.17: Yield curves for ^{18}O$(\alpha,n)^{21}$Ne obtained with enriched ^{18}O targets. Different levels of contamination are indicated by different peak shapes.

5.4 Background Correction Techniques

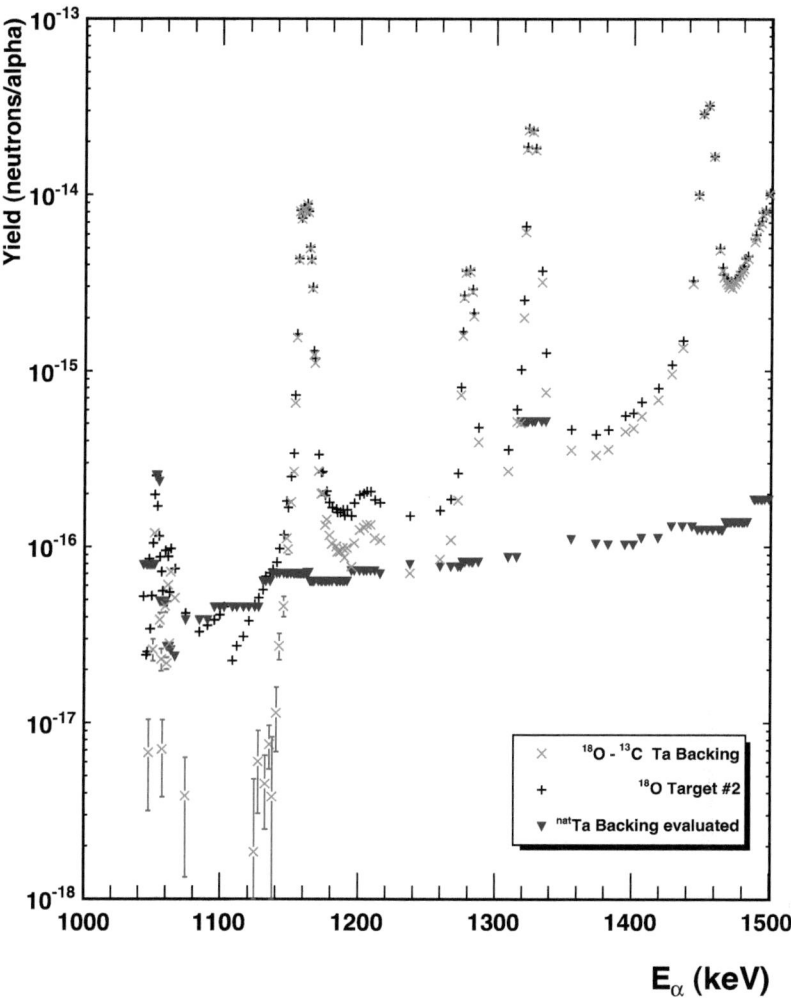

Figure 5.18: Shown are the yield curve of an ^{18}O target, the corresponding evaluated Ta background data and the corrected ^{18}O (^{18}O - ^{13}C) data.

Mg Background Target

Figure 5.19 shows the ^{24}Mg data compared to the data taken from a ^{25}Mg target. Both measurements show very similiar peak structures below an energy of $E_\alpha \leq 1900$ keV. Due to the low cross sections of the ^{25}Mg$(\alpha,n)^{28}$Si and ^{26}Mg$(\alpha,n)^{29}$Si reactions, the level of contamination has a stronger influence as for the ^{18}O$(\alpha,n)^{21}$Ne measurements.

The ^{24}Mg data are plotted with different scales in order to match different energy regions. The scaling factor of 1 allows to only match the region at 1053 keV while a scaling factor of 1.3 matches, e.g, the region at 1230 keV.

The ^{18}O$(\alpha,n)^{21}$Ne data are only influenced by the ^{13}C$(\alpha,n)^{16}$O background reaction, while the Mg data in general are influenced by at least two background reactions, ^{13}C$(\alpha,n)^{16}$O and ^{18}O$(\alpha,n)^{21}$Ne. Although each Mg target was produced and handled in the same manner, the levels of contamination varied considerably which prevented a successful correction with the routine employed in the case of ^{18}O$(\alpha,n)^{21}$Ne.

5.4 Background Correction Techniques

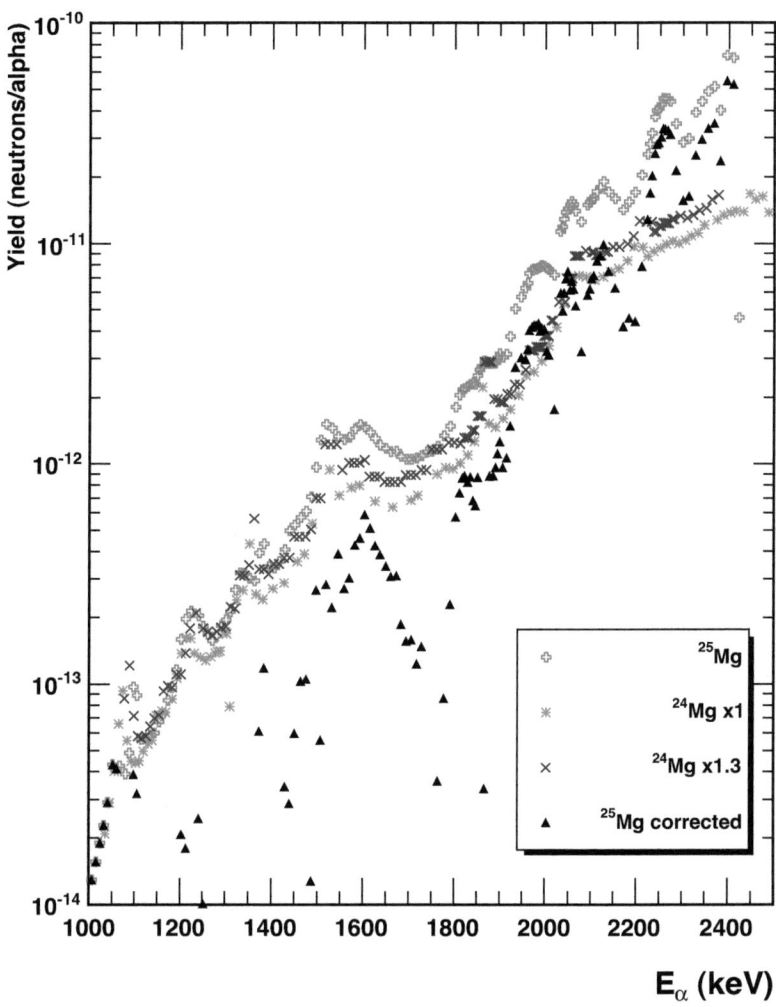

Figure 5.19: Yield data obtained with an enriched ^{25}Mg target, an enriched ^{24}Mg target (scaled by factors of 1 and 1.3, respectively) data and the corrected ^{25}Mg$(\alpha,n)^{28}$Si yield curve. For different scaling factors, the ^{24}Mg data can only match certain energy regions (see discussion in section 5.4.1).

5.4.2 Artifical Background Correction

Since a correction of the Mg production data with the background target measurement approach for $^{18}O(\alpha,n)^{21}Ne$ was not feasible, a different correction method was developed. This method involves the artificial introduction of background data into the experimental data set.

At first, peak structures within the uncorrected data were identified and then the cross section data sets for the identified background reaction acquired. For the $^{13}C(\alpha,n)^{16}O$ reaction the data of Harrissopulos et al. were used, while the $^{18}O(\alpha,n)^{21}Ne$ and $^{17}O(\alpha,n)^{20}Ne$ reaction data were taken from Denker and Bair et al.[8, 44, 98]. In order to apply the corrections, the cross section data of the background reactions were converted into experimental yields. The yield can be described as :

$$Y = \sigma \cdot \frac{\Delta}{\epsilon} \quad (5.12)$$

where Δ is the target thickness and ϵ the respective stopping power. By varying the target thickness and stopping power, it is possible to make assessments on the form of the contamination. This includes the geometrical location and contamination levels within the experimental setup.

To determine the geometrical location of the contamination the stopping power was varied. The variation was based on physical aspects and was not simply a parameter variation. For example, one can assume in principle that a possible ^{13}C contamination is located in the AuNiCu target backing of the Mg targets. This assumption is not unrealistic, considering the production processes of the AuNiCu backings. The backings are coated with Au by electrolysis and it is possible that AuCN and $Ni(CN)_2$ complexes are built into the Au layer during the electrolysis process[100]. An impinging α-particle interacting with a ^{13}C nucleus in the backing, inducing the $^{13}C(\alpha,n)^{16}O$ reaction will experience a different effective stopping power as compared to an α-particle reacting in a pure ^{13}C layer. Additionally, the thickness of the background source has to be taken into account. The previous example corresponds to a virtually an infinitely thick background source incorporated into the backing. This would lead to a different shape and magnitude of the yield curve[27]. This is also illustrated in figure 5.20.

The peak position of the background resonance serves as a parameter as well, since it points to the amount of energy an incoming α-particle loses before interaction with the impurity. As figure 5.15 shows the position of the $^{13}C(\alpha,n)^{16}O$ $E_r = 1053$ keV resonance peak can differ by several keV compared to the $^{13}C(\alpha,n)^{16}O$ cross section data. The energy shift can be observed as well for other resonances and therefore leaves no doubt that the peak is from the $^{13}C(\alpha,n)^{16}O$ reaction.

The energy shift corresponds to a thickness very similar to that of the corresponding Mg target. This indicates that the impurity is located between the Mg layer and the AuNiCu backing. A shift to higher energies would indicate a location behind the Mg layer, e.g. in the backing (as discussed before) or between backing and target material. Conversely, a shift to lower energies for example would indicate the impurity sitting on top of the Mg layer, as it is the case for the observed $^{18}O(\alpha,n)^{21}Ne$ resonances (see e.g. figure 5.21).

5.4 Background Correction Techniques

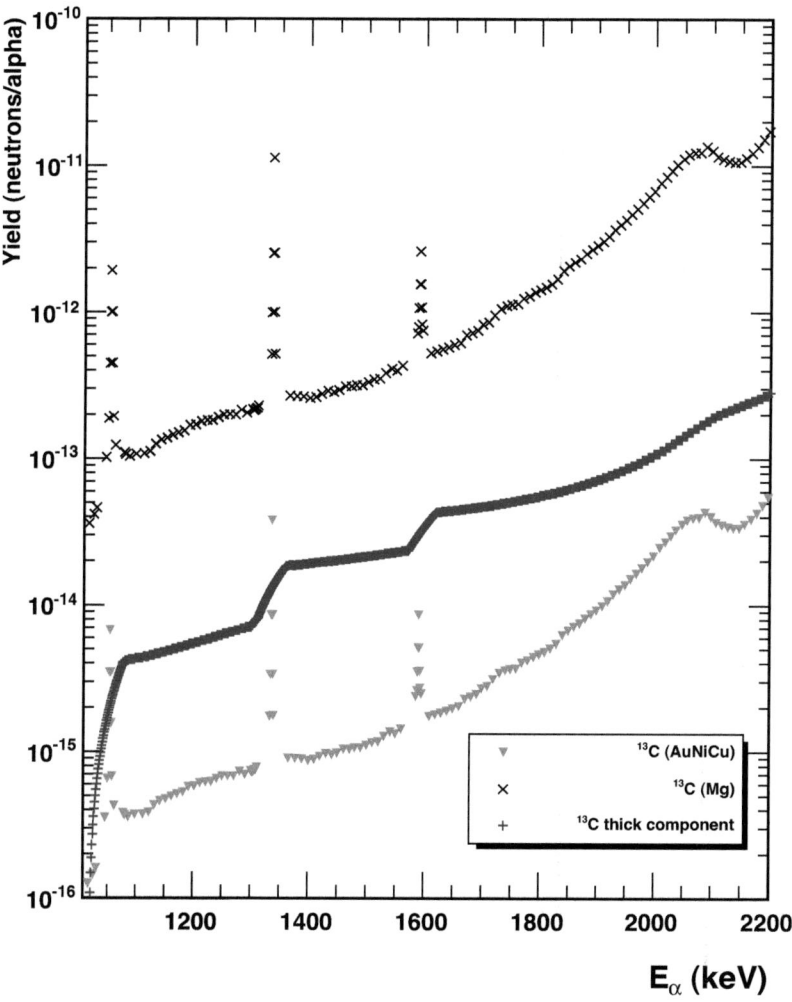

Figure 5.20: This plot illustrates the behaviour and magnitude for different types of ^{13}C impurities. If the ^{13}C is incorporated throughout the AuNiCu backing, the experimental data are influenced as displayed through the *^{13}C thick component* curve. *^{13}C (AuNiCu)* illustrates a thin impurity within the AuNiCu backing, while *^{13}C (Mg)* displays the ^{13}C background component for a thin impurity in the Mg layer.

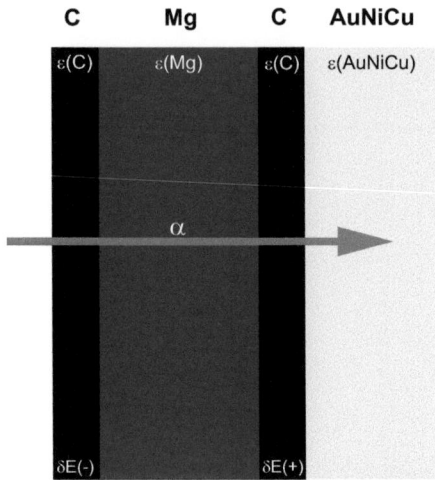

Figure 5.21: Schematic drawing of a Mg target. As an example two different C components, at different locations, within a target are shown. $\delta E(+)$ and $\delta E(-)$ stand for an energy shift to higher, respectively lower energy, while ϵ is the respective stopping power experienced by the impinging α-particle. The widths of the layers do not correspond to actual proportions.

In summary the method involves the following steps :

- Identify background reactions

  If possible, assign peak structures to know resonances of possible background reactions

- Calculate reaction yield for identified background reactions

  Cross section data reported in the literature are converted into reaction yield data, taking into account parameters such as efficiency, enrichment etc.

- Determine possible physical location of impurity

  Based on physical meaningful assumptions, vary stopping power and target thickness of possible background reactions to match actual experimental data.

- Implement energy shift

  Adjust the background reaction yield in energy and assign shifts according to location of impurity.

- Match data and correct

  Evaluate calculated reaction yield at energies given by experimental measurement. Correct experimental data by subtracting background reaction yield.

For each production target different levels and locations of contaminations were identified. Additionally, the number of contributing contaminations was determined. For example,

5.4 Background Correction Techniques

in figure 5.15 it is shown that the measurement not only included one, but two ^{13}C background components.

The reactions ^{17}O$(\alpha,n)^{20}$Ne was excluded as a major background contribution. This is due to the low natural abundance of the isotope and the relatively low cross section compared to the ^{18}O$(\alpha,n)^{21}$Ne reaction.

Corrections for the reaction ^{13}C$(\alpha,n)^{16}$O and ^{18}O$(\alpha,n)^{21}$Ne do not explain the resonant structure in the energy region of E_α = 1200 keV - 1300 keV. This resonant structure is observed in each Mg measurement. Wieland also reported this structure but did not give an explanation[45].

A search for possible (α,n) reactions occuring in this energy region led to the reaction ^{11}B$(\alpha,n)^{14}$N as the most likely missing background contribution[101]. Not only the region E_α = 1200 keV - 1300 keV, but also the region E_α = 1500 keV - 1600 keV shows obvious contributions from the ^{11}B$(\alpha,n)^{14}$N (see figure 5.22). The shape and location of the ^{11}B$(\alpha,n)^{14}$N resonance peaks in the yield curves indicates a, within uncertainties, consistent contamination of the Mg targets. This is probably due to an incorporation through the evaporation process.

The correction method explained above and including the ^{11}B$(\alpha,n)^{14}$N contribution has been applied especially to the Mg data and shows promising results. The detailed formalism for the correction method is described in Appendix C.

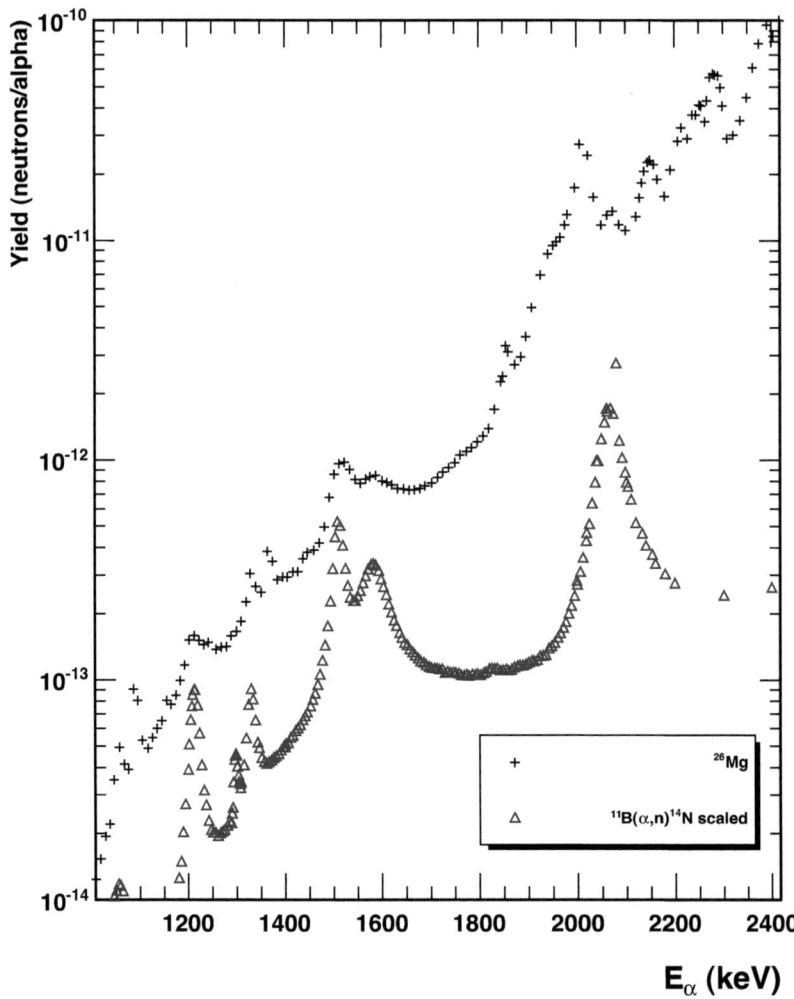

Figure 5.22: Yield data of a ^{26}Mg$(\alpha,n)^{29}$Si measurement and the cross section for the reaction ^{11}B$(\alpha,n)^{14}$N[101]. The cross section is scaled to emphasize the correlation between both data sets.

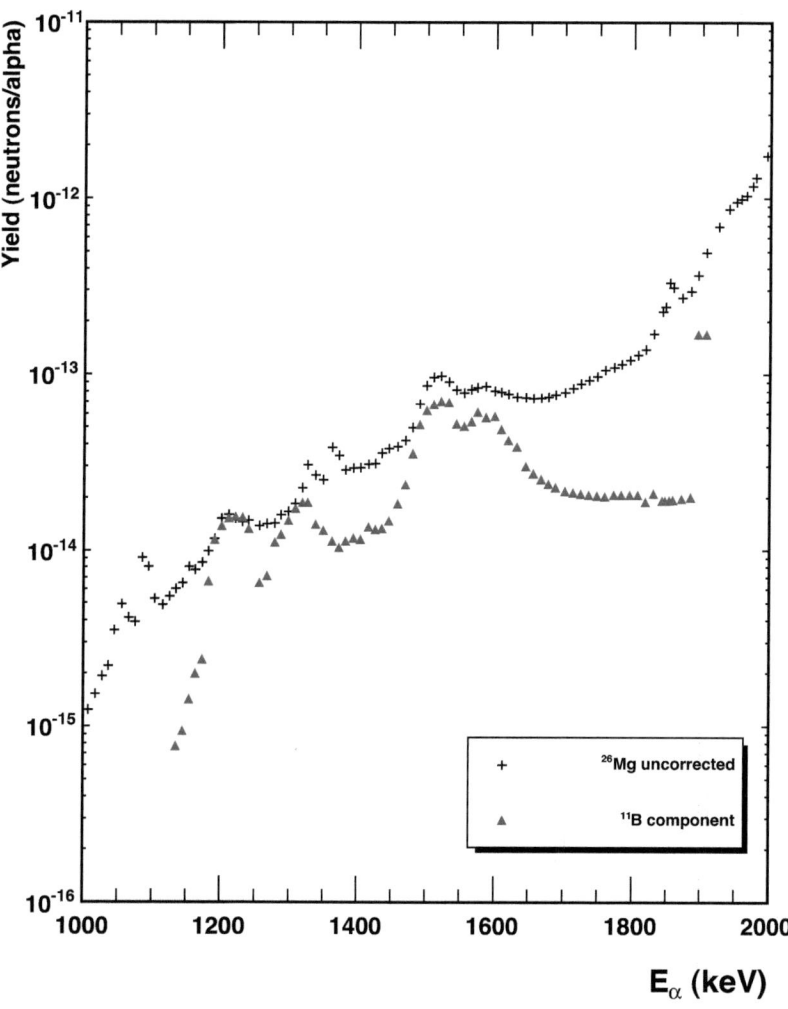

Figure 5.23: Yield curve of a ^{26}Mg$(\alpha,n)^{29}$Si measurement and the calculated background data for the ^{11}B$(\alpha,n)^{14}$N component. The ^{11}B$(\alpha,n)^{14}$N component is based on the cross section provided by Wang et al. and, in contrast to figure 5.22, is additionally matched for thickness, stopping power etc. [101].

5.5 ^{25}Mg(α,n)^{28}Si

The experimental data on ^{25}Mg(α,n)^{28}Si showed two ^{13}C, one ^{18}O and one ^{11}B background component. Below E_α = 1800 keV, the background does not allow a distinction between neutrons resulting from ^{25}Mg(α,n)^{28}Si or from one of the background reactions. As a consequence, upper limits only were calculated for this energy region (see figure 5.24).

For $E_\alpha \geq$ 1800 keV the experimental data clearly show resonant structures from which only the region around E_α = 1866 keV had to be excluded due to the observation of the ^{18}O(α,n)^{21}Ne(E_r= 1866 keV) resonance. The resonances and their according resonance parameters (resonance strength $\omega\gamma$, total width Γ and resonance energy E_r) were fitted and calculated. In table 5.4 they are listed and compared to the data reported by Wieland. The ambiguities in the Wieland data, such as a missing information on the background treatment, do not allow a quanititative comparison. Only the resonances at E_r = 1834, 2106, 2244 and 2338 keV could be confirmed in this work. An additional validation is given by the review of reported excited states (E_x) in the corresponding Nuclear Data Sheets (see table 5.3) [102, 103].

The derived resonance parameters and cross sections lead to an astrophysical S-factor which is up to an order of magnitude lower than the one reported by Wieland (see figure 5.26). In particular the calculated resonance strengths differ substantially (see table 5.4). This should have an immediate impact on the reaction rate.

States of ^{29}Si			This work	Wieland	NDS
E_α	E_{cm}	E_x			
2338	2015	13142	✓	✓	
2244	1934	13061	✓	✓	
2106	1816	12942	✓	✓	✓
2069	1783	12910		✓	
2035	1754	12881	✓		
1967	1695	12822	✓		
1944	1676	12802		✓	
1834	1581	12707	✓	✓	
1818	1567	12693	✓		✓
1784	1537	12665		✓	

Table 5.3: List of resolved resonance states for ^{25}Mg(α,n)^{28}Si, compared to the work of Wieland and others (NDS) [45, 103]. Note, that the resonances at E_α = 1967, 1944, 1834, 1818 overlap within errors (see also table 5.4).

5.5 ^{25}Mg(α,n)^{28}Si

This work [keV]					Wieland [keV]				
E_α	E_{cm}	E_x	Γ_{cm}	$\omega\gamma_{cm}$	E_α	E_{cm}	E_x	Γ_{cm}	$\omega\gamma_{cm}$
2338 ± 3	2015	13412	81.20 ± 2.74	9.48E-04 ± 2.29E-05	2334 ± 10	2012	13139	20 ± 4	25 ± 4
2244 ± 2	1934	13061	66.84 ± 7.83	6.80E-04 ± 7.00E-06	2245 ± 5	1935	13062	20 ± 5	20 ± 2
2106 ± 3	1816	12942	108.19 ± 3.5	3.80E-04 ± 1.04E-05	2116 ± 10	1824	12951		
					2069 ± 5	1784	12910		
2035 ± 2	1754	12880	44.37 ± 1.31	1.49E-04 ± 3.13E-06					
1967 ± 2	1695	12822	86.24 ± 1.26	1.31E-04 ± 1.55E-06	2006 ± 15	1729	12856	8± 3	2.6 ± 0.8
					1944 ± 15	1676	12802	4± 2	0.93± 0.4
					1896 ± 15	1634	12761	22 ± 10	0.42 ± 0.2
1834 ± 3	1581	12707	21.97 ± 3.87	1.30E-05 ± 1.00E-06	1837 ± 20	1584	12710	5± 2	0.27 ± 0.17
1818 ± 4	1567	12693	96.55 ± 3.42	3.28E-05 ± 1.30E-06					
					1784 ± 7	1538	12664	5± 3	0.25 ± 0.18

Table 5.4: ^{25}Mg(α,n)^{28}Si resonance parameters from Wieland compared to the parameters obtained in this work[45].

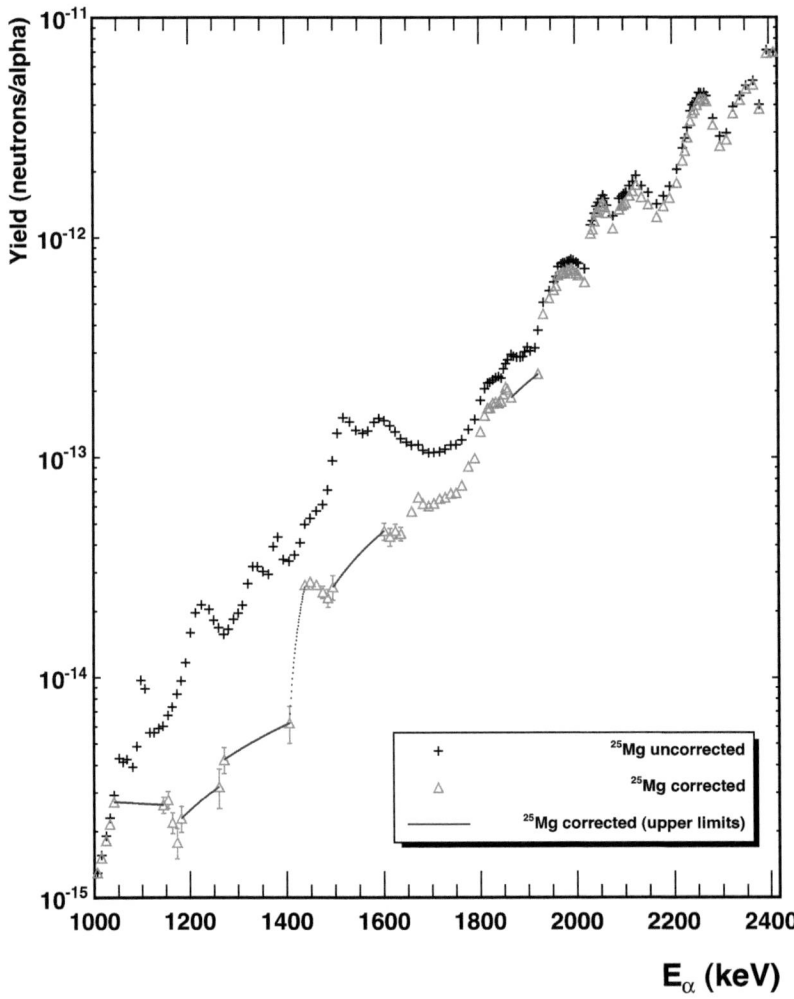

Figure 5.24: Illustrated are the measured experimental yields, the yields after background correction and upper limits. The upper limits are especially located in energy regions where the background can not be separated from the data set.

5.5 ^{25}Mg(α,n)^{28}Si

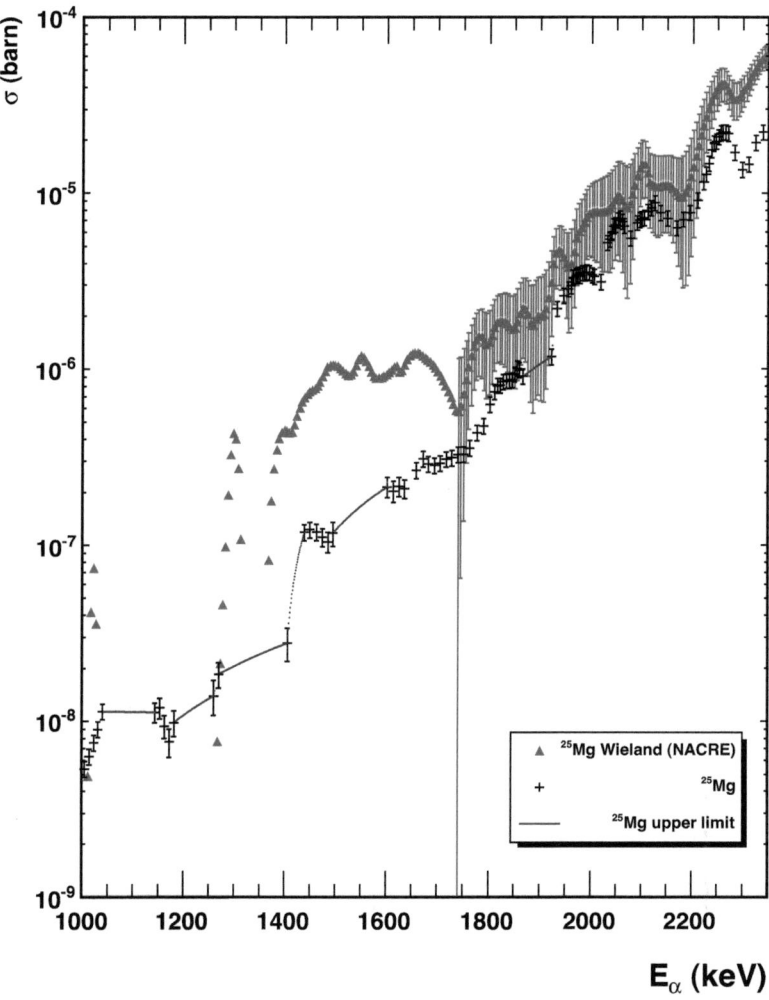

Figure 5.25: Illustrated is the cross section for ^{25}Mg(α,n)^{28}Si derived from the present data, compared to the data provided by Wieland[45].

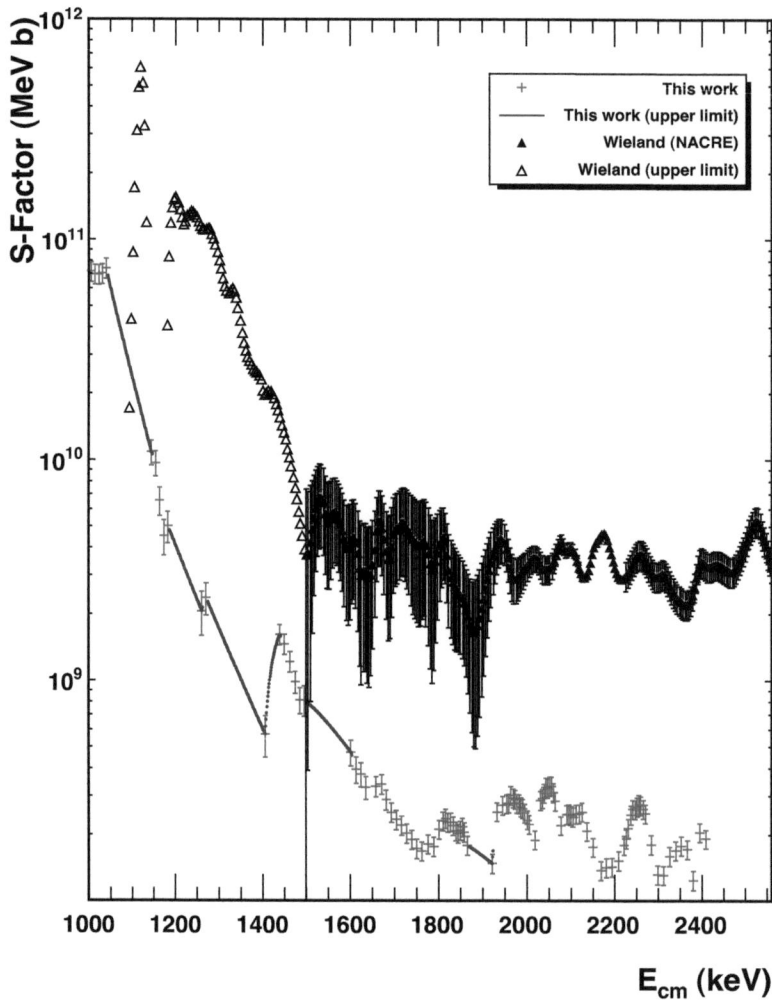

Figure 5.26: Illustrated is the S-factor for ^{25}Mg$(\alpha,n)^{28}$Si obtained, compared to the NACRE data. The NACRE data below 1700 keV are listed as upper limits and show no error bars[45, 57].

5.6 ^{26}Mg(α,n)^{29}Si

The results for the ^{26}Mg(α,n)^{29}Si reaction are similar to the ones obtained for ^{25}Mg(α,n)^{28}Si. A separation from the background reactions below $E_\alpha = 1750$ keV is not possible. After the background correction, the peak structures below $E_\alpha = 1750$ keV completely disappear (see figure 5.27). This, again, raises serious concerns regarding the reported resonances and cross sections by Wieland[45].

Five resonances for the reaction ^{26}Mg(α,n)^{29}Si were determined. These were seen at least in one of the previous reports (see table 5.5). The fitted resonance parameters again differ substantially from the ones previously reported by Wieland (see table 5.6). Several resonance (or excited) states reported before by Küchler and NDS could not be resolved as well[43, 102, 103]. As a result the astrophysical S-factor was determined to be significantly lower as previously reported (see figure 5.28).

States of ^{30}Si			This work	Wieland	Küchler	NDS
E_α	E_{cm}	E_x				
2397	2077	12720		✓	✓	
2373	2057	12700	✓			✓
2290	1985	12628		✓	✓	
2267	1965	12608	✓			
2233	1935	12579	✓	✓		
2215	1920	12563			✓	
2146	1860	12503			✓	
2135	1850	12494	✓			
2123	1840	12483		✓		
2077	1800	12443			✓	
2054	1780	12423	✓			
2030	1759	12402				✓
2010	1742	12385		✓	✓	
1996	1730	12373	✓			
1957	1696	12339	✓	✓		
1909	1654	12298		✓		
1846	1600	12243		✓		
1800	1560	12203			✓	✓
1781	1544	12187		✓		

Table 5.5: List of resolved resonance states for ^{26}Mg(α,n)^{29}Si, compared to the work of Wieland, Küchler and others (NDS) [43, 45, 103]. Note, that, analogous to the case of ^{25}Mg(α,n)^{28}Si, resonances overlap within errors (see also table 5.6).

This work [keV]					Wieland [keV]				
E_α	E_{cm}	E_x	Γ_{cm}	$\omega\gamma_{cm}$	E_α	E_{cm}	E_x	Γ_{cm}	$\omega\gamma_{cm}$
2373 ± 2	2057	12700	73.84 ± 1.45	1.10E-06 ± 1.00E-05	2397 ± 6	2077	12720	34 ± 10	90 ± 20
2267 ± 2	1965	12608	32.76 ± 0.68	3.63E-07 ± 6.90E-06	2290 ± 5	1985	12628	22 ± 8	27 ± 14
2233 ± 3	1935	12579	92.12 ± 1.75	5.70E-07 ± 1.82E-05	2237 ± 6	1939	12582		
2135 ± 2	1850	12494	52.96 ± 1.01	1.90E-07 ± 4.31E-06					
2054 ± 3	1780	12423	98.30 ± 4.81	1.85E-07 ± 1.39E-05	2123 ± 10	1840	12483	40 ± 19	20 ± 14
					2010 ± 5	1742	12385	10 ± 4	10 ± 4
1996 ± 2	1730	12373	43.42 ± 0.78	1.66E-09 ± 2.96E-06	1953 ± 15	1693	12302		
1957 ± 3	1696	12339	73.06 ± 0.96	1.76E-09 ± 3.15E-06	1909 ± 10	1655	12298		
					1846 ± 10	1600	12243		
					1781 ± 15	1544	1287	5 ± 3	0.08 ± 0.04

Table 5.6: Obtained resonance parameters for ^{26}Mg(α,n)^{29}Si, compared to the parameters reported by Wieland[45].

5.6 ^{26}Mg(α,n)^{29}Si

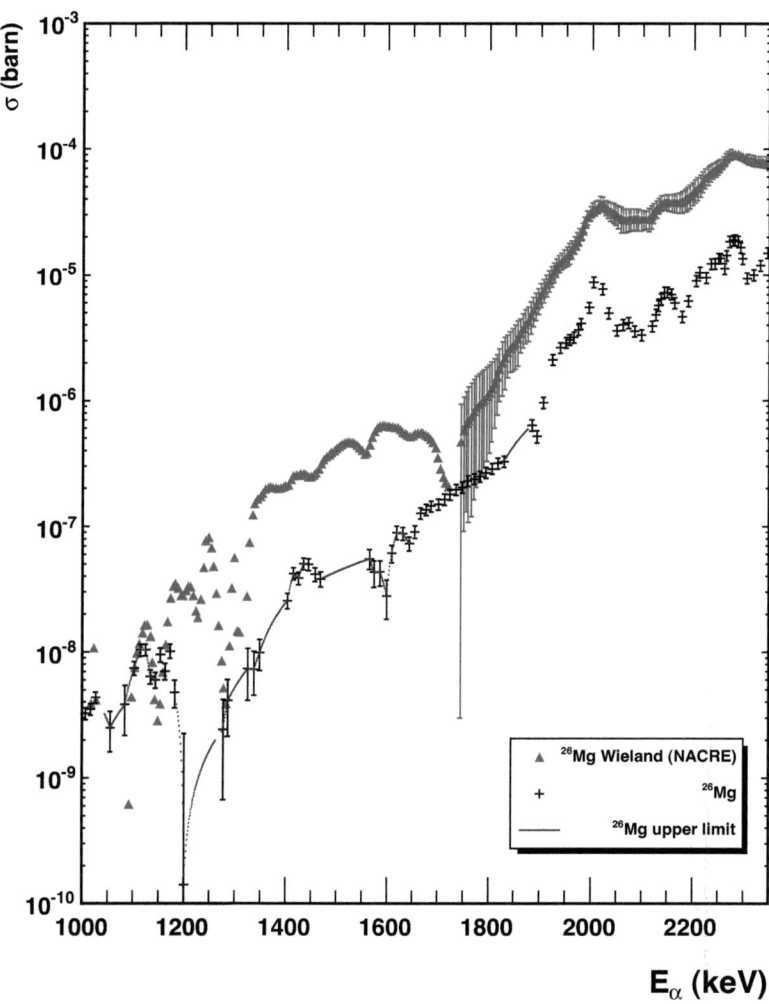

Figure 5.27: Shown are the cross sections for ^{26}Mg(α,n)^{29}Si obtained in this work, compared to the cross section data reported by NACRE. It has to be noted, that the NACRE data are dominated by the data achieved of Wieland[45, 57]. Below 1700 keV they are listed as upper limits and show no error bars.

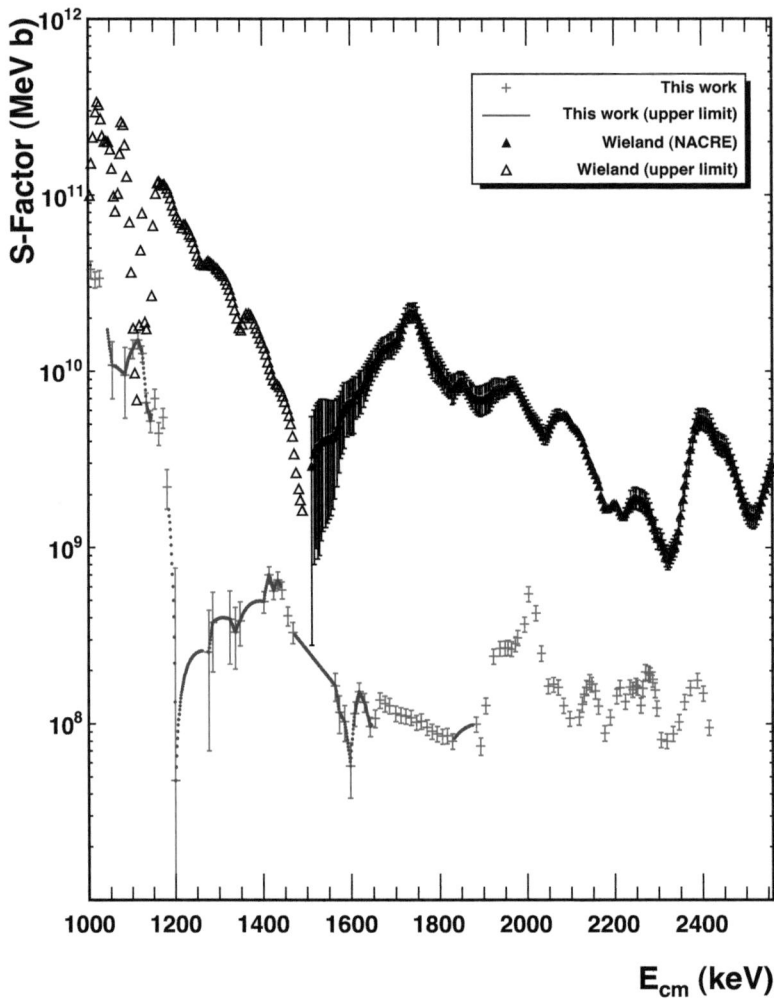

Figure 5.28: Illustrated is the S-factor for ^{26}Mg(α,n)^{29}Si obtained in this work, in comparison to the NACRE data. The NACRE data below 1700 keV are listed as upper limits and therefore show no error bars.

5.7 ^{18}O(α,n)^{21}Ne

For the ^{18}O(α,n)^{21}Ne experiment, only the ^{13}C(α,n)^{16}O reaction was found as a background component. The resonances reported by Denker (respectively NACRE) have been confirmed by the present experimental data[8, 57]. The resonances below 1050 keV could not be confirmed with the yield measurements.

The resonance parameters derived from the present data represent the first independent validation of the results from Bair and from Denker[8, 44]. In particular the achieved resonance strengths $\omega\gamma_{cm}$ are close to the ones reported by Denker, which is important for the calculation of the reaction rate. For the first time resonance parameters also for the $E_r = 1817$ keV resonance are reported.

However, the cross sections differ from the ones reported in the NACRE compilation[57]. The difference is up to a factor of 4, and an energy shift of about 8 keV is observed (see figures 5.29 and 5.30). Denker's work reveals two possible sources for this ambiguity. For one, the determination of the detection efficiency was performed with MCNP only and a few experimental data points. These data points did not necessarily agree well with the MCNP simulations, in particular since the MCNP simulations from the present work had to be adjusted (scaled) to match the experimental results. A second possibility is the treatment of the stopping power. Denker used a differential gas target system which requires a more sophisticated determination of the stopping power.

Nevertheless, the astrophysical S-factor shows a very good agreement between the present data and previously obtained results (see figure 5.31).

This work [keV]				NDS [keV]			
E_α	E_{cm}	E_x	Γ_{cm}	E_α	E_{cm}	E_x	Γ_{cm}
2466 ± 2	2017	11685	8 ± 3	2467 ± 6	2018	11686	9
2340 ± 2	1915	11583	20 ± 4	2335 ± 6	1910	11578	18
2199 ± 2	1799	11467	8 ± 3	2195 ± 4	1796	11464	≤ 3
2165 ± 2	1771	11439	57 ± 8	2160 ± 10	1767	11435	48
1959 ± 3	1603	11271	4 ± 1	1955 ± 6	1600	11268	12
1867 ± 3	1528	11196	6 ± 1	1864 ± 4	1525	11193	7
1817 ± 2	1487	11155	14 ± 1	1787 ± 6	1462	11130	≤ 5
1665 ± 2	1362	11030	17 ± 3	1667 ± 7	1364	11032	≤10
1531 ± 4	1253	10921	16 ± 1	1531 ± 4	1253	10921	24
1452 ± 3	1188	10856	5 ± 1	1453 ± 4	1189	10857	6
1324 ± 4	1083	10751	2 ± 1	1321 ± 4	1081	10749	6
1277 ± 2	1045	10713	3 ± 2	1269 ± 5	1038	10706	≤ 10
1157 ± 3	947	10615	2 ± 1	1159 ± 4	948	10616	6
1064 ± 3	870	10538	8 ± 2	1079 ± 18	883	10551	

Table 5.7: Resonance parameters obtained for the reaction ^{18}O(α,n)^{21}Ne compared to the results from Nuclear Data Sheets (NDS)[104]. The results from NDS for example, include the data from Bair et al.[44]. No results for the resonance strengths are given by NDS (or included references).

This work [keV]					Denker [keV]				
E_α	E_{cm}	E_x	Γ_{cm}	$\omega\gamma_{cm}$	E_α	E_{cm}	E_x	Γ_{cm}	$\omega\gamma_{cm}$
2466 ± 2	2017	11685	8 ± 3	1.667E+00 ± 1.513E-02					
2340 ± 2	1915	11583	20 ± 4	3.109E-01 ± 1.612E-03					
2199 ± 2	1799	11467	8 ± 3	3.890E-01 ± 3.840E-03					
2165 ± 2	1771	11439	57 ± 8	3.891E-01 ± 4.622E-03					
1959 ± 3	1603	11271	4 ± 1	1.538E-02 ± 2.089E-03	1958 ± 4	1602	11270	6 ± 2	2.200E-02 ± 1.949E-03
1867 ± 3	1528	11196	6 ± 1	9.211E-02 ± 1.127E-02	1866 ± 4	1527	11195	6 ± 3	1.580E-01 ± 1.640E-02
					1839 ± 7	1505	11173		
1817 ± 2	1487	11155	14 ± 1	2.462E-03 ± 1.516E-04	1817 ± 7	1487	11155	14 ± 5	
1665 ± 2	1362	11030	17 ± 3	1.578E-03 ± 4.394E-04	1660 ± 4	1358	11026	≤5	1.590E-04 ± 1.400E-05
1531 ± 4	1253	10921	16 ± 1	2.652E-02 ± 1.537E-03	1526 ± 4	1249	10917	25 ± 3	4.170E-02 ± 4.060E-03
1452 ± 3	1188	10856	5 ± 1	2.786E-03 ± 2.786E-04	1443 ± 3	1181	10849	≤4	2.490E-03 ± 2.200E-04
1324 ± 4	1083	10751	2 ± 1	1.148E-03 ± 2.399E-04	1317 ± 4	1078	10746	4	1.940E-03 ± 1.700E-04
1277 ± 2	1045	10713	3 ± 2	2.362E-04 ± 7.345E-05	1270 ± 3	1039	10707	≤3	2.460E-04 ± 2.100E-05
1157 ± 3	947	10615	2 ± 1	4.554E-04 ± 1.178E-04	1152 ± 3	943	10611	≤3	6.400E-04 ± 5.500E-05
					1127 ± 7	922	10590	20 ± 3	
1064 ± 3	870	10538	8 ± 2	7.585E-06 ± 3.661E-06	1057 ± 3	865	10533	≤5	3.230E-06 ± 2.700E-07

Table 5.8: Resonance parameters obtained for the reaction $^{18}O(\alpha,n)^{21}Ne$ in this work, compared to the results of Denker[8].

5.7 ^{18}O(α,n)^{21}Ne

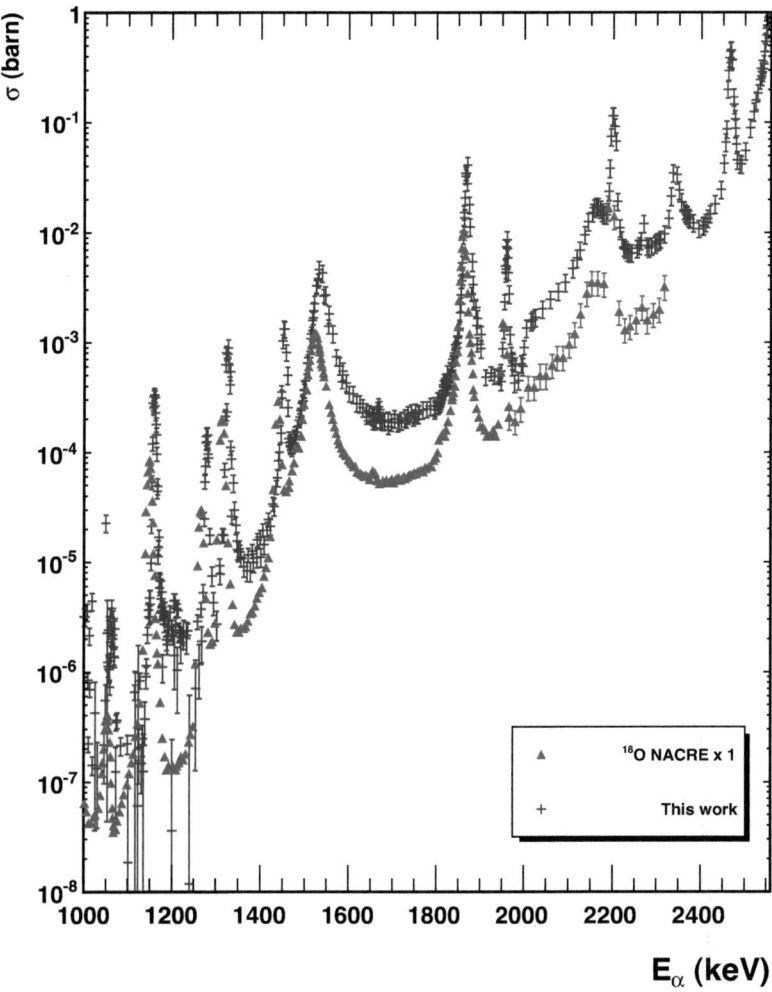

Figure 5.29: Illustrated is the cross section for ^{18}O(α,n)^{21}Ne, compared to the NACRE data, which includes the data obtained by Bair and those obtained by Denker[8, 44, 57]. Note, that an energy shift and a scaling difference between both data sets exists.

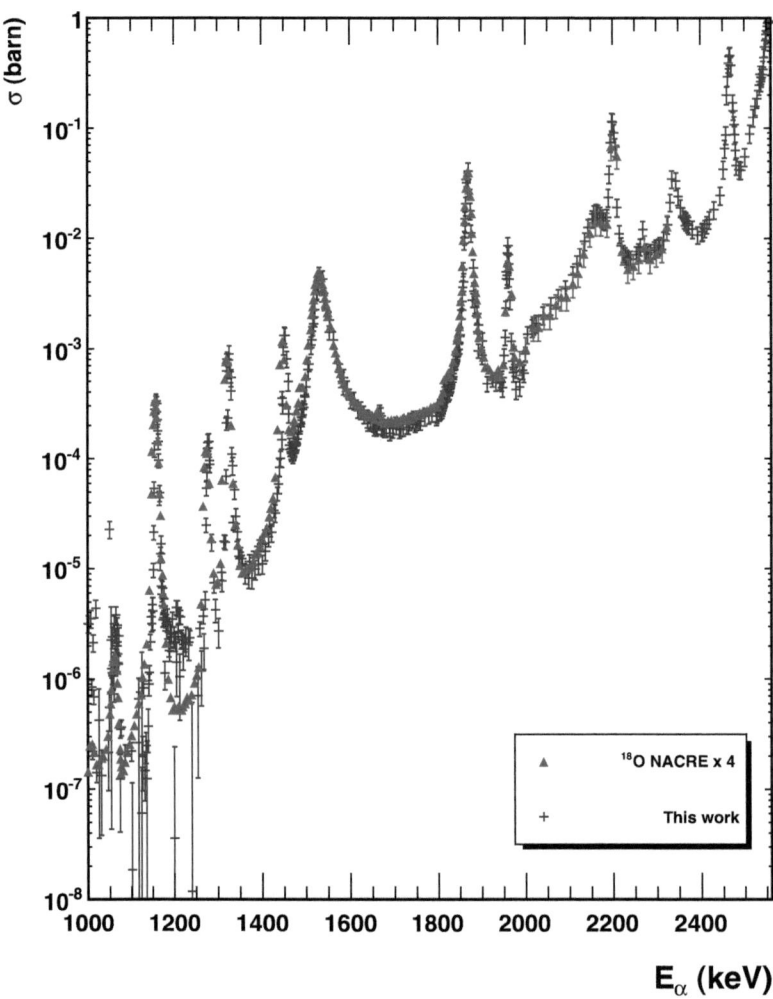

Figure 5.30: Shifted and scaled NACRE data set for $^{18}O(\alpha,n)^{21}Ne$ in comparison to the experimental data set obtained in the present work.

5.7 ^{18}O$(\alpha,n)^{21}$Ne

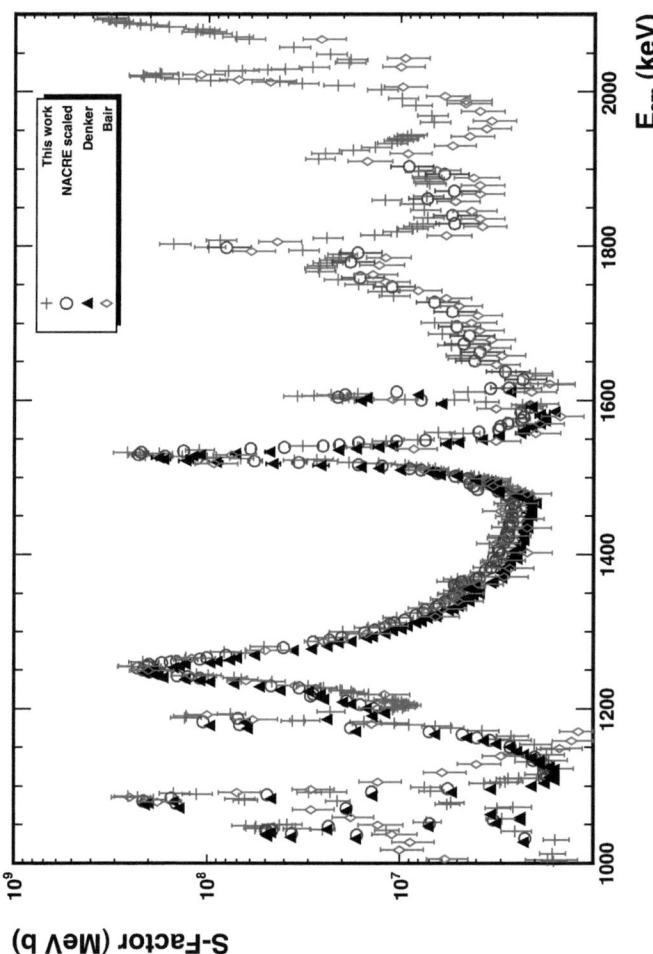

Figure 5.31: The astrophysical S-factor from this work compared to the scaled and un-scaled NACRE data.

5.8 Results from Ring Ratio Calculations

The ring ratio for each reaction was utilized to determine the energy of the neutrons detected. This is especially useful to (a) determine the neutron group detected and (b) investigate effects on the detection efficiency. As mentioned earlier, different neutron groups have different energies which affect the detection efficiency. Different ring ratios would indicate if an observed resonance belongs to one neutron group exclusively or if a mixing of neutron groups is occurring. A shortcoming is that the ring ratios represent a *mean* neutron energy $\bar{E}_n$, if mixing of different neutron groups occurs.

Figures 5.32 and 5.33 show the measured ring ratios for ^{25}Mg(α,n)^{28}Si and ^{26}Mg(α,n)^{29}Si. Unfortunately, a clear distinction of the ring ratio data from the reaction of interest from the background data (especially the resonances) was not possible. For ^{18}O(α,n)^{21}Ne the situation is more favorable (see figure 5.34), since all resonances can be observed. In the following, the analysis method will be discussed for the case of ^{18}O(α,n)^{21}Ne.

First one has to review the nuclear structure information important for the reaction mechanism. For (α,n) reactions the compound nucleus mechanism can applied. This assumes that the target nucleus and the incoming α-particle form a compound nucleus, which decays via the emission of a neutron. The important parameters are listed in table 5.9.

Parameter	^{18}O(α,n)^{21}Ne	^{25}Mg(α,n)^{28}Si	^{26}Mg(α,n)^{29}Si
Q_α	9668	10698	10643
S_n	10364	8044	10609
E_{thres}	696	0	0
Q_0	-696	2654	34
Q_1	-1047	874	-1236
Q_2	-2442	-1966	-1996

Table 5.9: Nuclear structure parameters important for the compound nucleus mechanism from NDS[102, 103, 104].

The final nucleus of the reaction ^{18}O(α,n)^{21}Ne has three states within the experimental range that can be populated : The ground state at 0 keV, the first excited state at 351 keV and the 2nd excited state at 1746 keV. To cause the reaction to occur, the α-particle has to impinge on the ^{18}O nucleus with an energy greater than the threshold energy E_{thres} = 696 keV (E_α = 856 keV). This implies that resonances occuring below a laboratory resonance energy of :

$$E_r = (E_{thres} + 351) * \frac{M+m}{M} = 1277 keV \qquad (5.13)$$

exclusively populate the ground state of ^{21}Ne (note, that M+m/m = 18+4/18). Consequently, these resonances exclusively consist of neutrons from the n_0 group, since the energy is not high enough to populate the first excited state of ^{21}Ne.

5.8 Results from Ring Ratio Calculations

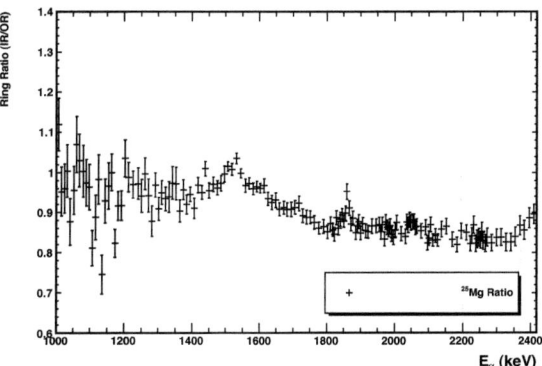

Figure 5.32: Illustrated is the ring ratio obtained for the ^{25}Mg$(\alpha,n)^{28}$Si measurement of an enriched ^{25}Mg target.

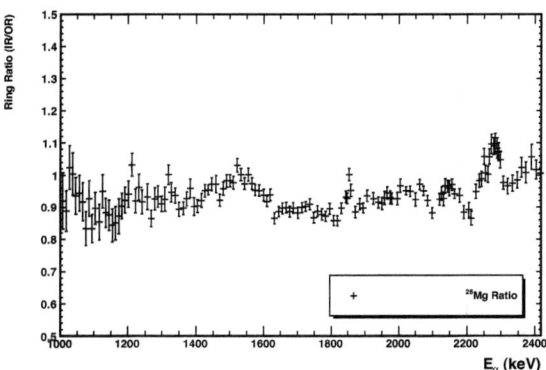

Figure 5.33: Illustrated is the ring ratio obtained for the ^{26}Mg$(\alpha,n)^{29}$Si measurement of an enriched ^{26}Mg target.

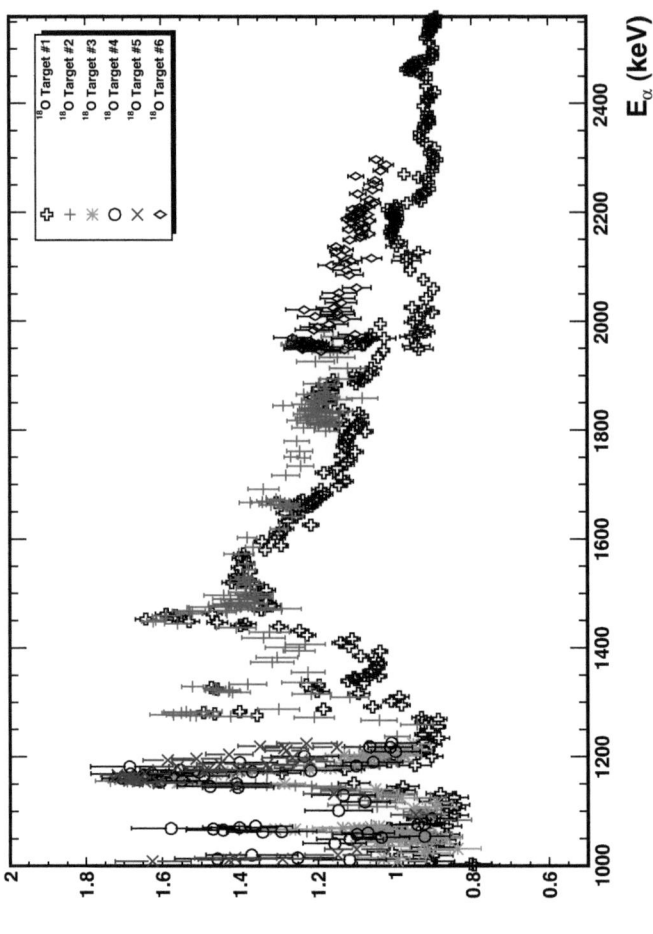

Figure 5.34: Yield curve of the ring ratio for the $^{18}O(\alpha,n)^{21}Ne$ reaction obtained with various enriched ^{18}O targets.

5.8 Results from Ring Ratio Calculations

One can estimate the mixing ratios of the neutron groups for the different resonances. The present data yield a ring ratio of 1.62 for the resonance at $E_r = 1157$ keV (see also table 5.10). If x_i denotes the mixing ratio of the ith neutron group (with $\sum x_i = 1$), one can note for the $E_r = 1157$ keV resonance:

$$R(1157 keV) = 1.62 = x_0 \cdot E_{n_0} + x_1 \cdot E_{n_1} \qquad (5.14)$$

with $E_r = 1157$ keV as an exclusive n_0-group resonance, equation 5.13 becomes simply :

$$E_n(1.62) = 1 \cdot E_{n_0} + 0 \cdot E_{n_1} = 206 keV \qquad (5.15)$$

A calculation of the neutron energy $((1157 \cdot 0.82) - E_{thres} = 252 keV)$ results within the sensitivity of the ring ratio.

E_r	$E_{n_0}^{calc}$	$\bar{E}_n^{data}$	R
1008	130		1.74*
1064	176	240	1.58*
1157	252	204	1.62
1277	351	372	1.46
1324	389	333	1.49
1452	494	353	1.47
1531	559	189	1.64
1665	669	667	1.31
1817	793	1250	1.14
1839	811	1028	1.19
1867	834	1079	1.18
1959	910	793	1.26
2165	1079	2079	1
2199	1106	1963	1.02
2270	1160	2316	0.97†
2340	1222	2682	0.94
2466	1325	2360	0.97

Table 5.10: Calculated neutron energies for a 100 % branching into the n_0 group ($E_{n_0}^{calc}$), the observed ratios and calculated mean neutron energies. The resonances marked with * could not be separated sufficiently, while the resonance marked with † was not observed.

The resonance at $E_r = 1277$ keV is the first resonance that is open to the n_1 channel. If the n_0 channel is exclusively populated, the neutron energy should be 351 keV. The present data for the $E_r = 1277$ keV resonance shows :

$$E_n(1.46) = 372 keV \qquad (5.16)$$

which allows to claim a pure n_0 resonance. In table 5.10 a list of the ring ratios for the different resonances and the corresponding mean neutron energies is given. $E_{n_0}^{calc}$ represent the neutron energy, that one should observe, if the resonance were an exclusive n_0 resonance. If mixing with the n_1-group occurs, the mean neutron energy should be higher :

$$\bar{E}_n(R) = x_0 \cdot E_{n_0} + x_1 \cdot E_{n_1} > x_0 \cdot E_{n_0} \quad (5.17)$$

since $E_{n_1} = E_{n_0}$ - 350 keV.
However, in some cases the data show also :

$$E_{n_0}^{calc} > \bar{E}_n^{data} \quad (5.18)$$

which indicates the occurrence of interferences between resonances. This leads to an increase in the detected mean neutron energy. For example, the resonances in the energy region 1800 - 1900 keV fulfill the relation of equation 5.16 and interfere, while isolated resonances (e.g. E_r = 1966 keV) fulfill equation 5.15.

Table 5.11 shows the estimated mixing ratios for resonances, which fulfilled equation 5.15. It has to be noted that these values are *estimates* and probably are not very accurate. The uncertainty is introduced by a missing background reduction for the ring ratio data. In order to reliably correct for background contributions, one would have to measure the possible background reactions with exactly the same detector setup. The ring ratios measured would then allow a reliable correction. Additionally, the sensitivity of the ring ratio with respect to the neutron energy is not strong enough to yield more accurate results. Still, such measurements would probably also result in better data sets for the reactions ^{25}Mg(α,n)^{28}Si and ^{26}Mg(α,n)^{29}Si.

To raise confidence in the achieved mixing ratios, one could use the J^π assignments of the involved nuclei and possibly estimate the transition probabilities. Experimentally, it may be possible to design a more sensitive detector geometry for the specific reaction of interest. However, these considerations and their implementation exceed the scope of this thesis.

In addition to the resonances identified in the experimental yield data, a resonance was found at approximately E_r = 2270 keV. This would correspond to E_x = 11528 keV in the compound nucleus, in agreement with an excited state at E_x= 11530 found previously through the ^{20}Ne(t,p)^{22}Ne reaction[104].

In any case, the detection efficiency is not affected significantly by the occurrence of different neutron groups and is not very sensitive to the ring ratio, as previously shown in figure 5.8.

E_r	x_0	x_1
1008	1	
1064	1	
1157	1	
1277	1	
1324	0.84	0.16
1452	0.6	0.4
1531	0	1
1665	1	0
1959	0.66	0.34

Table 5.11: Calculated mixing ratios for ^{18}O(α,n)^{21}Ne resonances. The resonances not listed here do not allow a conclusive calculation due to the occurence of the n_2 group or interference effects with other resonances.

5.9 Reaction Rates

As already discussed in section 2.4, the reaction rate determines the outcome of the network calculations, which will lead to an impact assessment for stellar nucleosynthesis. Nuclear structure effects and uncertainties in the experimental data entail the use of assumptions for the determination of the reaction rate.

In general, the reaction rate is dependent on the reaction mechanism. For (α,n) reactions the mechanism involves the formation of a compound nucleus. As a consequence, the occurrence of resonant behaviour is strongly favored, which results in a resonant component of the reaction rate (see also 2.3):

$$\langle \sigma v \rangle_r = \left(\frac{2\pi}{\mu kT}\right)^{3/2} \hbar^2 \sum_i (\omega\gamma)_i exp\left(-\frac{E_i}{kT}\right) f \qquad (5.19)$$

where i denotes the specific resonance observed. For ^{25}Mg(α,n)^{28}Si and ^{26}Mg(α,n)^{29}Si this is the situation in the energy range E_α = 1800 keV. As illustrated previously, the experimental data do not allow an accurate determination of the cross sections down to lower astrophysical energies, however. For these energy regions the cross section, or respectively S-factor, was assumed to be constant at 1.36 · 10^8 MeVb for ^{25}Mg(α,n)^{28}Si and 7.97 · 10^8 MeVb for ^{26}Mg(α,n)^{29}Si. The corresponding reaction rate can then be expressed as :

$$\langle \sigma v \rangle_{sfactor} = \left(\frac{2}{\mu}\right)^{1/2} \frac{\Delta}{(kT)^{3/2}} S(E_0)(1 + \frac{5}{12\tau}) exp(-\tau) \qquad (5.20)$$

with Δ as the effective width of the relevant energy window and τ being defined as :

$$\tau = \frac{3E_0}{kT} \gg 1 \qquad (5.21)$$

This constitutes then a second component for the reaction rate[27].

For energy regions completely outside the experimental range investigated in this thesis, the cross section data sets were complemented. By data taken from the NACRE compilation the total reaction rate becomes therefore :

$$\langle \sigma v \rangle_{total} = \langle \sigma v \rangle_r + \langle \sigma v \rangle_{sfactor} + \langle \sigma v \rangle_{NACRE} \qquad (5.22)$$

including the energy ranges and resonances not investigated in this thesis ($\langle \sigma v \rangle_{NACRE}$)[27, 57].

Due to the substantial background corrections and resulting lower resonance strengths, the reaction rates for the reactions ^{25}Mg(α,n)^{28}Si and ^{26}Mg(α,n)^{29}Si obtained in this work are up to several orders of magnitude lower than those previously reported. This is illustrated in figures 5.35 and 5.36, where the different calculated components and the data from literature are shown. The differences between each component illustrate the importance of accurate resonance information from experimental data sets.

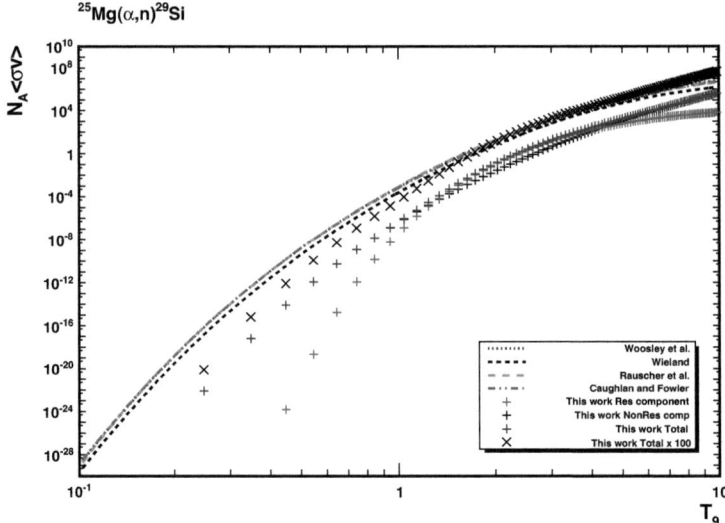

Figure 5.35: Comparison to previous results of the obtained reaction rate as a function of temperature for ^{25}Mg$(\alpha,n)^{28}$Si. The results from previous authors are represented by analytical fits[45, 47, 48, 57, 105]. To illustrate the effect of the different components from this work, the analytical fits were excluded from this figure. At $T_9 = 1.5$ the reaction rate for ^{25}Mg$(\alpha,n)^{28}$Si is by a factor of $1.2 \cdot 10^{-2}$ lower as compared to the reaction rate reported by Wieland. The reaction rate is multiplied by Avogadro's constant and displayed in units of cm^3 mol^{-1} s^{-1}.

5.9 Reaction Rates 111

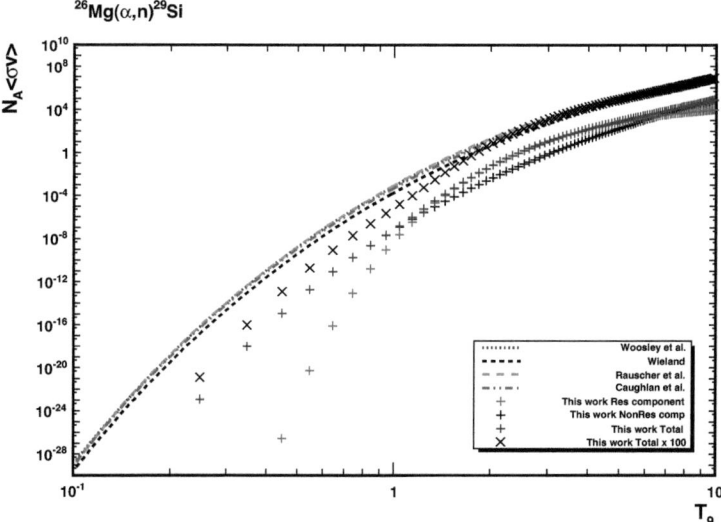

Figure 5.36: Displayed are the different reaction rate components as a function of temperature for the reaction ^{26}Mg(α,n)^{29}Si and the results from previous authors, analogous to figure 5.35. At $T_9 = 1.5$ the reaction rate for ^{26}Mg(α,n)^{29}Si is by a factor of $7 \cdot 10^{-3}$ lower as compared to the reaction rate reported by Wieland.

The situation for the ^{18}O(α,n)^{21}Ne the reaction rate is illustrated in figure 5.37. The calculation of the reaction rate resulted in an increase by a factor of about 4 due to the higher measured cross sections. The shift in resonance energies does not have an essential effect.

A comparison of the reaction rate to the reaction rates of ^{18}O(α,γ)^{22}Ne by Dababneh et al. showed a shift for the temperature at which the (α,n) channel is dominating over the (α,γ) channel (see figure 5.38 and table 5.12). The shift is caused by the new experimental data available through Dababneh et al. and the reaction rate from this work. Denker, for example, was only able to compare to older results on ^{18}O(α,γ)^{22}Ne [8, 58]. The shift indicates a release of additional neutron fluxes and an impact on steller nucleosynthesis at an earlier stage of stellar evolution.

Authors	Temperature [T_9]
Denker et al.	0.6
Denker et al., Dababneh et al.	0.62
This work, Dababneh et al.	0.52

Table 5.12: List of temperatures at which the ^{18}O(α,n)^{21}Ne reaction starts to dominate over the ^{18}O(α,γ)^{22}Ne reaction[8, 58].

5.9 Reaction Rates

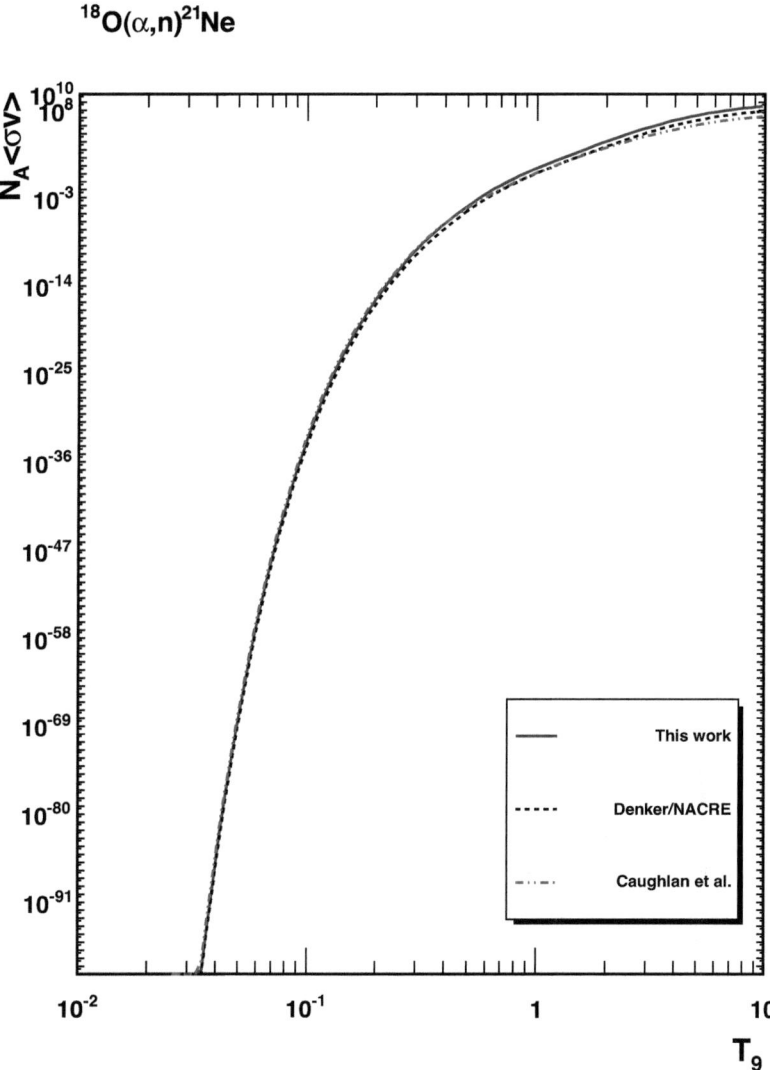

Figure 5.37: Comparison of the reaction rate obtained for the reaction $^{18}O(\alpha,n)^{21}Ne$ and previous results[8, 48, 57].

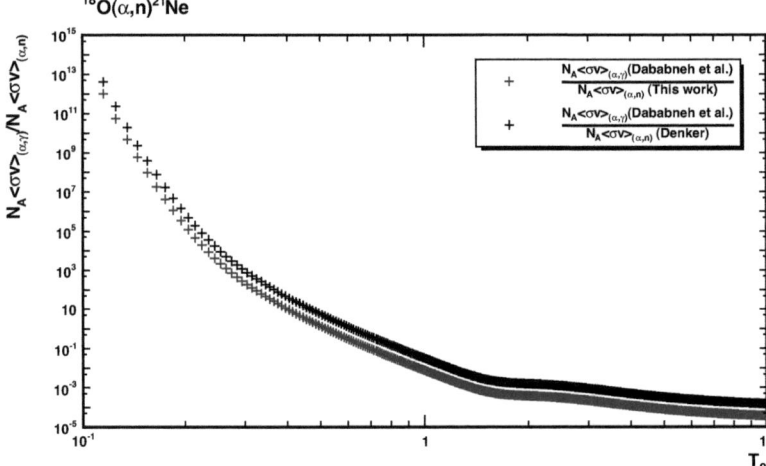

Figure 5.38: Illustrated is the ratio of the obtained reaction rate for ^{18}O$(\alpha,n)^{21}$Ne to the reaction rate of the (α,γ) channel. The ratio of the reaction rates include the data from Denker and Dababneh et al.[8, 58].

Chapter 6
Impact Studies

The impact of the previously obtained reactions rates can be quantitavely assessed by implementing the reaction rates into network calculations (see section 2.4). The PPN NUGRID post-processing code calculates the abundance distribution for each isotope involved in the nucleosynthesis processes[30].

In the context of this thesis, it is important to note that the results from PPN are based on so-called one-zone calculations. One-zone calculations provide the abundances (and their variations) for one given phase of the stellar evolution. ^{25}Mg$(\alpha,n)^{28}$Si and ^{26}Mg$(\alpha,n)^{29}$Si become relevant during Ne burning conditions, while in the case of ^{18}O$(\alpha,n)^{21}$Ne, He burning conditions have to be considered. This constrains the impact of the reactions to specific aspects of stellar nucleosynthesis.

6.1 ^{25}Mg(α,n)^{28}Si & ^{26}Mg(α,n)^{29}Si

The obtained reaction rates of ^{25}Mg(α,n)^{28}Si and ^{26}Mg(α,n)^{29}Si are significantly different from those provided by NACRE[57]. Considering the large uncertainties still associated with the reaction rates, a sensitivity study was performed. This study contained the variation of the reactions rates up to two orders of magnitude in comparison to the reaction rates recommended by NACRE.

To observe the impact of these variations, specific isotopes were chosen, whose abundances were estimated to be influenced by the change in reaction rates. Apart from the abundances of the Mg and Si isotopes, whose abundances are influenced directly, the isotopic abundances of titanium, strontium, zirconium, molybdenum and barium were chosen as observables for the impact. The abundances of these isotopes are influenced indirectly through the varying neutron exposures.

To account for different initial stellar masses and the Ne-burning conditions in the core as well as in the convective shell, the calculations were performed at $T_9 = 1.5$ and 1.7. For sake of simplicity, the matter density was kept constant at 10^6 cm^{-3}. The end of the burning phase was conventionally set to the time at which the ^{20}Ne abundance would have dropped to approximately 1%. This would typically be the case after one year for $T_9 = 1.5$, respectively 1.4 days for $T_9 = 1.7$. Notice that under such conditions, convective Ne-shell burning is terminated by the supernova explosion, while ^{20}Ne is still available[10]. Nucleosynthesis in the pre-explosive Ne-shell is mostly affected by the reactions during the initial burning phase, when ^{20}Ne is more abundant. As the ^{20}Ne abundance decreases in the shell, the amounts of α-particles, neutrons and protons rapidly drop as well. For this reason, the results of our impact study may be applied to both Ne-burning conditions, in the core (complete ^{20}Ne exhaustion) and in the shell (partial ^{20}Ne exhaustion).

The initial seed abundances for the calculations are taken from the C-burning ashes of a of 25M$_\odot$ and solar-like metallicity star[23].

The calculations provide the abundance distribution at the end of the Ne shell not only for stable but also for unstable isotopes which can contribute to the abundances of the stable isotopes through their β-decays. This additional component is called *radiogenic component*. The half-life of the contributing isotope has to be long enough to contribute during the condensation phase of the ejected material, e.g. observed in form of the SiC X grains (see secton 2.5).

In figure 6.1, the neutron density is plotted as a function of time during neon burning for the different cases considered in our study. The maximum neutron density is obtained in the first phases of Ne burning followed by a rapid decrease. At the beginning, the high ^{20}Ne abundance provides a large flux of α-particles via the ^{20}Ne(γ,α)^{16}O reaction.

These α-particles may then be captured by ^{25}Mg and ^{26}Mg to produce neutrons. The burning of ^{25}Mg and ^{26}Mg via α-capture and the decrease of ^{20}Ne cause the neutron density to decrease with time. The different neutron peak densities illustrated in figure 6.1 show that ^{25}Mg(α,n)^{28}Si and ^{26}Mg(α,n)^{29}Si provide about 50% of the total neutron density as evident by comparing the different cases when their respective reaction rates are decreased, kept the same or set to zero.

In figures 6.2 - 6.7, the obtained abundance distributions and their variation at the end of Ne shell burning are shown. In most cases, the radiogenic contribution to the stable isotopes is negligible or small compared to the original abundances of the stable species. For a small number of considered cases, however the radiogenic contribution is indeed significant, and does affect the final isotopic distribution. The most significant ones are

6.1 ^{25}Mg(α,n)^{28}Si & ^{26}Mg(α,n)^{29}Si

listed in table 6.1 and plotted in figures 6.2 - 6.7 for different feeding amounts as well.

Decay chain	Longest lived isotope	$t_{1/2}$
$^{46-50}$Sc $\to$ $^{46-50}$Ti	^{44}Sc	84 d
^{86}Rb $\to$ ^{86}Sr	^{86}Rb	19 d
^{90,91}Sr $\to$ ^{90,91}Y $\to$ ^{90,91}Zr	^{90}Sr	29 a
^{94}Nb $\to$ ^{94}Mo	^{94}Nb	20300 a
^{95}Zr $\to$ ^{95}Nb $\to$ ^{95}Mo	^{95}Zr	64 d
$^{134-138}$Cs $\to$ $^{134-138}$Ba	^{135}Cs	$2 \cdot 10^6$ a

Table 6.1: List of decay chains affecting significantly the final abundances of the daughter nuclei, for the elements considered in this discussion (Ti, Sr, Zr, Mo, Ba). The longest lived isotopes in each chain and their respective half-lives are listed as well.

Particular cases, that require detailed discussion, are the nuclei ^{94}Nb and ^{135}Cs.
^{135}Cs is accumulated during Ne-burning, due to its long half life ($t_{1/2} = 2$ Ma). Two aspects have to be taken into account : (1) At the end of Ne-burning, ^{135}Cs is abundant enough to possibly contribute to the ^{135}Ba abundance. To have a noticeable effect (at least 1% feeding) on the ^{135}Ba abundance, a feeding time of at least 30000 years was determined. Since SiC X grains form within a few years, only a marginal fraction of ^{135}Cs will decay to ^{135}Ba before grain formation occurs[106, 107]. (2) Due to its chemical properties, Cs does not condense in grains and therefore does not contribute to the abundance of Ba in SiC X grains after grain formation.
^{94}Nb is a long-lived isotope ($t_{1/2} = 2 \cdot 10^4$ a), which is accumulated during Ne-burning as well and contributes through its decay to the ^{94}Mo abundance. The difference to the case of ^{135}Cs is, that ^{94}Nb condenses into silicon carbide grains and contributes to the ^{94}Mo abundance *after* grain formation [108, 109] (see figures 6.3, 6.5 and 6.7).

^{25}Mg(α,n)^{28}Si Figures 6.2 and 6.3 show the abundance variation for ^{25}Mg(α,n)^{28}Si and rate variations by factors of 0.1 and 0.01 relative to NACRE, at $T_9 = 1.5$ and 1.7[57]. With the reduction of the reaction rate, the abundance of ^{25}Mg is increased up to 100%. The direct reaction product, ^{28}Si, is reduced to 20% percent.
The ^{26}Mg and ^{29}Si abundances, for example, are reduced as well. This can be assigned to less neutrons being available for neutron captures on ^{25}Mg and subsequently reducing the occurrence of the ^{25}Mg(n,γ)^{26}Mg(α,n)^{29}Si reaction chain.
The calculations for $T_9 = 1.5$ (see figure 6.2) imply significantly decreased abundances for most of the neutron capture species. Indeed, as a consequence of the reduced ^{25}Mg(α,n)^{28}Si reaction rate, a lower amount of neutrons produced can be observed (see figure 6.1).
Only ^{90}Zr, ^{94}Mo,^{98}Mo and ^{134}Ba are increased in their abundances compared to the NACRE reaction rate. Note that the increased abundances are also a result of the radiogenic component, especially for ^{90}Zr. As for ^{134}Ba, in the NACRE case it is largely bypassed, as the high neutron densities favour neutron capture compared to β-decay at the branching points ^{133}Xe and ^{134}Cs. Therefore, in our case ^{134}Ba is less destroyed as compared to the NACRE reaction rate.
For $T_9 = 1.7$ (see figure 6.3), most of the abundances for isotopes close to the valley of stability are enhanced, while for more neutron-rich species the abundances are decreased. Indeed, the increase in abundances, in particular for ^{94}Mo, is up to orders of magnitude

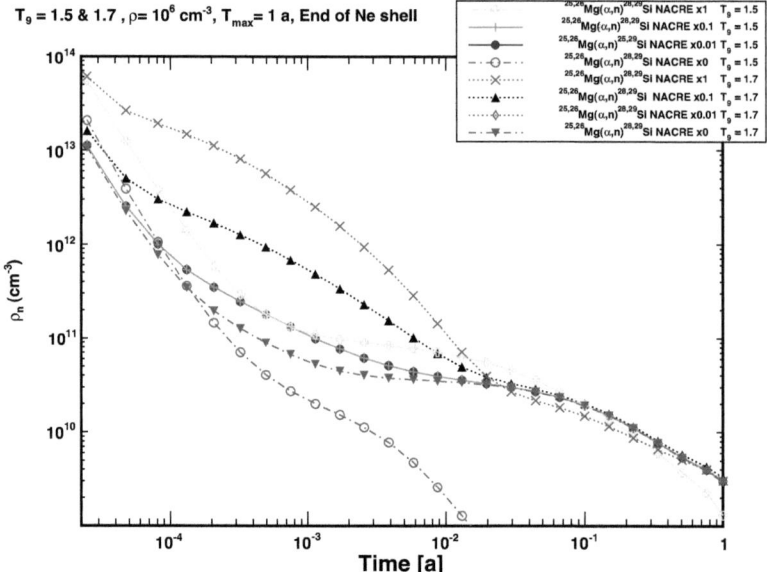

Figure 6.1: Neutron density as a function of time for $T_9 = 1.5$ & 1.7. For $T_9 = 1.7$ and after 1.4 days ($3.8 \cdot 10^{-3}$ a) the Ne burning phase stops due to the exhaustion of ^{20}Ne.

6.1 ^{25}Mg(α,n)^{28}Si & ^{26}Mg(α,n)^{29}Si

higher (depending on the radiogenic feeding) as compared to the case of $T_9 = 1.5$. This is caused by the reduced amount of neutrons (caused by the lower reaction rate) available for neutron captures as compared to the abundances calculated for the NACRE reaction rate.

^{26}Mg(α,n)^{29}Si The present calculations show a higher sensitivity to the production of neutrons by ^{26}Mg(α,n)^{29}Si in the Ne shell as compared to ^{25}Mg(α,n)^{28}Si. For example, at $T_9 = 1.5$ the abundance of ^{29}Si is more reduced with the lower reaction rate on ^{26}Mg(α,n)^{29}Si than the ^{28}Si abundance for the ^{25}Mg(α,n)^{28}Si case. Additionally, up to 65% less Sr is produced compared to about 45% less for ^{25}Mg(α,n)^{28}Si (see figures 6.4 and 6.5). In comparison to ^{25}Mg(α,n)^{28}Si, the abundances of the isotopes are quantitatively more affected by ^{26}Mg(α,n)^{29}Si, while the general trends are matched by both reactions.

The change in trends between $T_9 = 1.5$ and $T_9 = 1.7$ is the same for both reactions. ^{26}Mg(α,n)^{29}Si has a stronger influence on the abundance of the direct reaction products and the remaining Mg abundances.

The quantitative influence of both reactions on the abundance distribution of s-process element is similar. When adjusting the reaction rates of both reactions at the same time the effects on the abundance distributions become quantitatively stronger (see figures 6.6 and 6.7). The different behaviour between $T_9 = 1.5$ and $T_9 = 1.7$ is due to the different neutron density regimes found.

For $T_9 = 1.7$ an additional feature has to be taken into account. The observation of an increased abundance of for example the p-only nuclide ^{92}Mo indicates the significance of photodesintegrations (e.g. (γ,n) reactions). As the amount of neutrons becomes lower through the decreased reaction rates, the equilibrium between (n,γ) and (γ,n) reactions is shifted towards the destruction channel (γ,n). Additionally for (γ,n) reactions, one can estimate the time scale by the inverse of the reaction rate (or half-life, respectively).

Especially for the phase after ^{20}Ne is exhausted, no neutrons can be supplied to further fuel the (α,n) reactions.

Mo isotope	Time scale T_9 =1.5 [a]	Time scale T_9 =1.7 [a]
92	2.60E+15	2.06E+10
93	6.60E+02	3.40E-01
94	1.50E+06	1.31E+02
95	2.40E+00	2.40E-03
96	2.56E+04	5.21E+00
97	2.85E-02	4.68E-05
98	4.09E+02	1.32E-01
99	3.25E-05	1.33E-07
100	4.23E+01	1.80E-02

Table 6.2: Listed are the time scales for (γ,n) reactions on the Mo isotopes for T_9 =1.5 and T_9 =1.7. The times scales, in general, are several orders of magnitude smaller for higher temperature regimes.

Comparison to SiC X grains Presolar grains from massive stars form in the SN ejecta on a time scale of a few years after the SN explosion[106, 107]. In this phase, the SN ejecta are affected by large mixing effects between different zones of the star, where carbon rich material of the He shell is exposed and polluted by matter being mixed in from other regions of the star, including material from the Ne shell.

As Ne shell material is further processed through the SN explosion, the heavier nuclei previously discussed are affected by photodisintegration processes. However, in some models, the convective Ne shell is mixed with the convective C shell during the final day(s), thus causing a significant material exchange between both zones[110]. Under such conditions, isotopic signatures of the pre-explosive Ne-shell, together with C shell material, may be ejected almost unchanged through the SN explosion.

A clear disentanglement of the pre-explosive Ne-shell component from presolar SiC grains is not possible through the course of this work, but a comparison of Ne shell signatures to presolar grain data could qualitatively lead to implications for the origin of the presolar grain signatures, i.e. whether they may contain significant amounts of Ne shell material. To compare the obtained results with isotopic ratios in SiC X grains, the calculated abundance ratios were normalized to solar abundance ratios. For the example of the barium isotopes results are given in the form :

$$\delta\left(\frac{^{i}Ba}{^{136}Ba}\right) = \left\{\frac{\frac{X_i}{X_{136}}}{\left[\frac{x_i}{x_{136}}\right]_\odot} - 1\right\} * 1000 \qquad (6.1)$$

The results were normalized to the solar abundances given by Anders and Grevesse and compared to measurements on SiC X grains (see figures E.3 - E.10) [111, 112, 113].

The obtained results in comparison to Ti data from SiC X grains do not show an agreement (see figures E.1 - E.2). To some extent at least, this is due to missing information, e.g., from explosive nucleosynthesis and radiogenic components. These are not included since only Ne shell material is considered here.

In figure E.3, the isotopic signatures for Sr measured by Pellin et al. are compared to the obtained results at $T_9 = 1.5$[112]. Clearly, the calculations do not match the production of the isotope ^{88}Sr. However, for $T_9 = 1.7$, the measured isotopic signature seems to be better reproduced by the present calculations. This suggests that the material condensed in the grains measured by Pellin et al. was exposed to higher neutron densities than explored here.

Similar to the Sr abundances distribution, the calculations match the SiC X data for Zr (see figures E.5 - E.6) at $T_9 = 1.7$ better than to the results calculated for $T_9 = 1.5$.

The Mo signature in SiC grains gives a more robust information. Indeed, SiC X grains are characterized by a Mo distribution with an overproduction of ^{95}Mo, ^{97}Mo and a slight (or no) overproduction of ^{100}Mo. For example Meyer et al. claimed, that such a peculiar distribution may be reproduced by a neutron burst (starting from a solar abundance seed, previously being exposed to a weak neutron fluence) at the bottom of the He shell, in coincidence with the SN shock passage, indicating that at least for Mo a dominant nucleosynthesis signature in presolar grains can be generated in this event[114]. In figures E.7 - E.8, the results from the performed calculations are compared to SiC X data. Note, that for $T_9 = 1.7$ Ne shell burning reproduces also the SiC X data for ^{92}Mo relatively well, while in the case of $T_9 = 1.5$ no ^{92}Mo is produced.

6.1 ^{25}Mg(α,n)^{28}Si & ^{26}Mg(α,n)^{29}Si

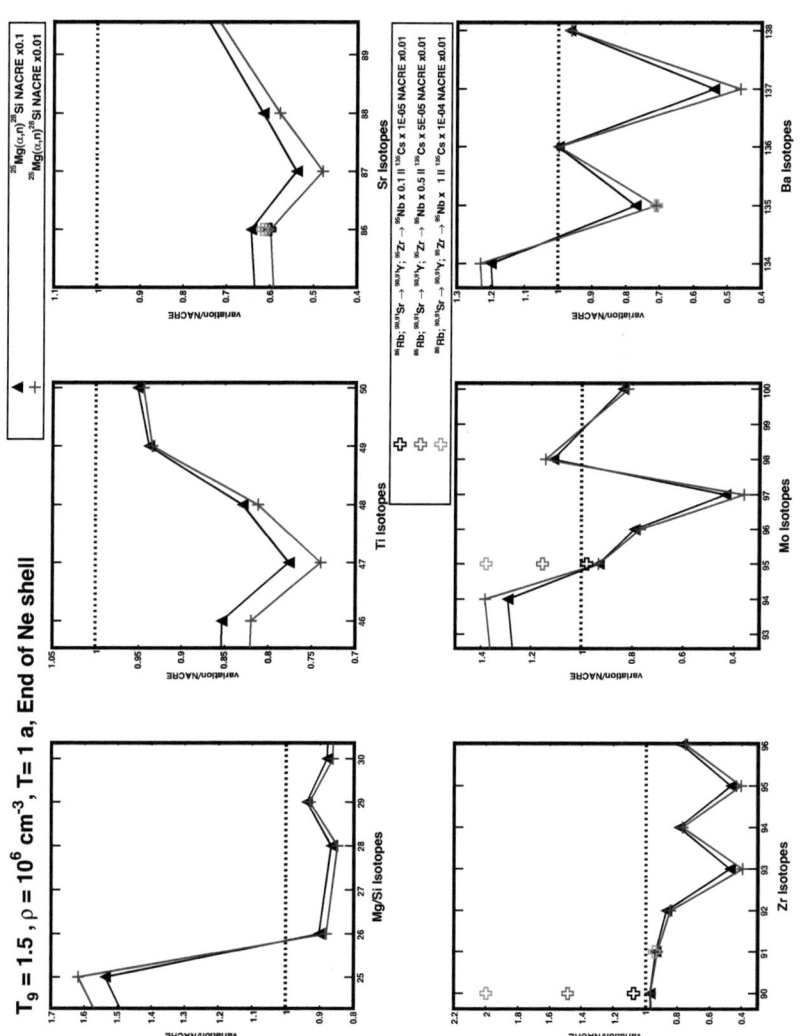

Figure 6.2: Illustrated are the isotopic abundance variations for a reduced reaction rate of ^{25}Mg(α,n)^{28}Si at $T_9 = 1.5$. The reaction rate is reduced by factors of 10 and 100 relative to NACRE[57]. Possible contributions through radiogenic components are shown for the most significant cases as well. The different factors (e.g. x 0.5) show the amount of decayed nuclei of the mother nuclide.

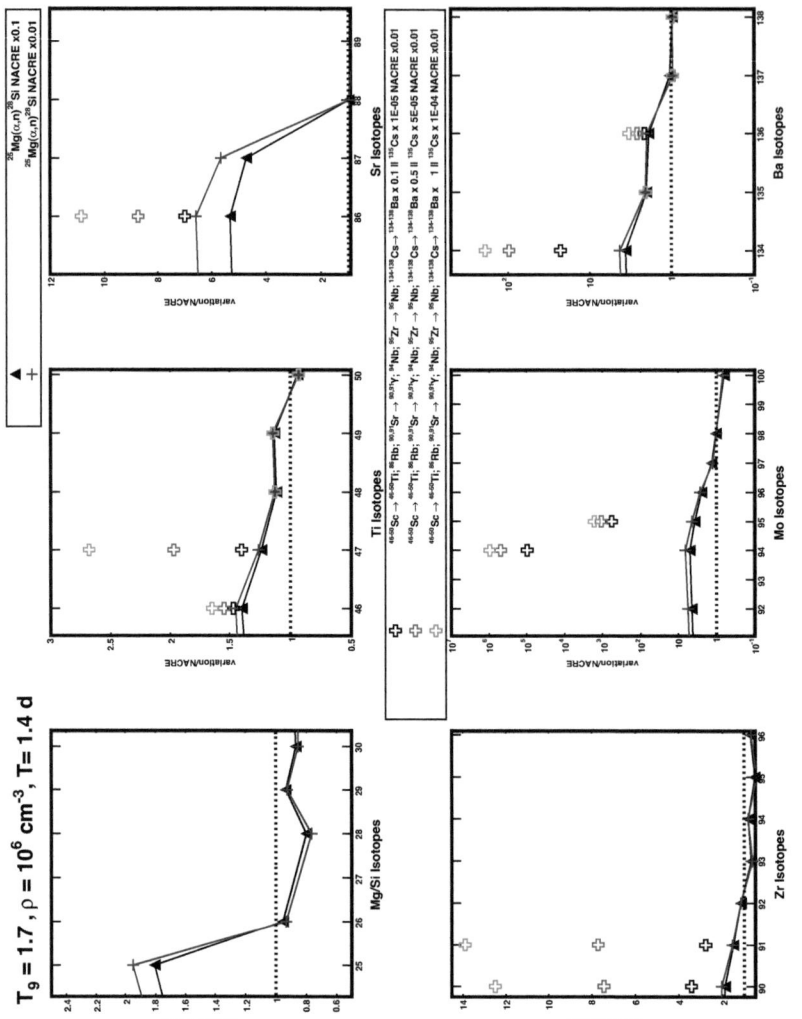

Figure 6.3: Illustrated are the isotopic abundance variations for a reduced reaction rate of $^{25}\text{Mg}(\alpha,n)^{28}\text{Si}$ at $T_9 = 1.7$ after 1.4 days of burning time, analogous to figure 6.2.

6.1 ^{25}Mg(α,n)^{28}Si & ^{26}Mg(α,n)^{29}Si

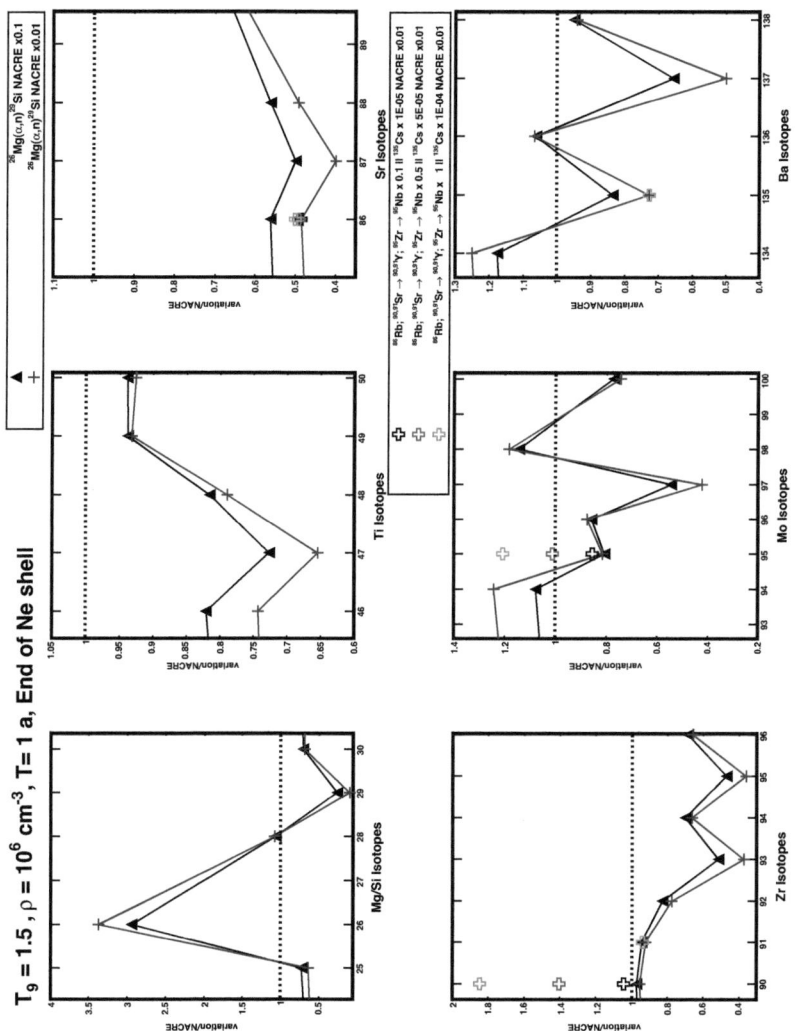

Figure 6.4: Illustrated are the isotopic abundance variations for a reduced reaction rate of ^{26}Mg(α,n)^{29}Si at $T_9 = 1.5$, analogous to figure 6.2.

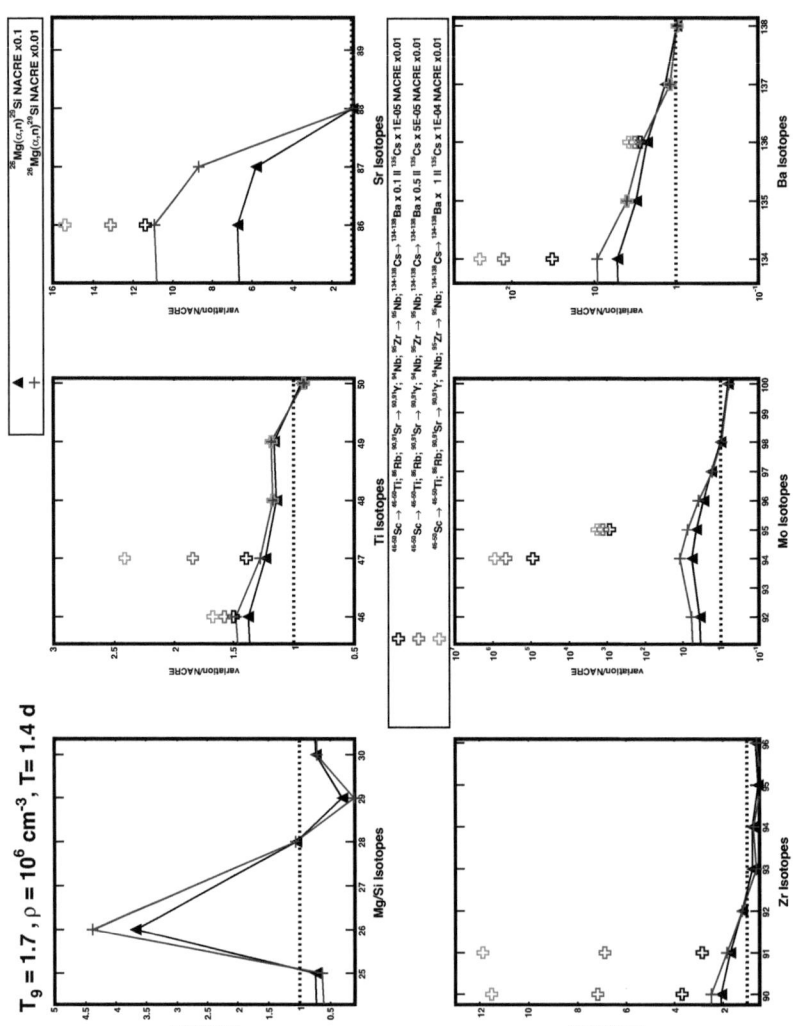

Figure 6.5: Illustrated are the isotopic abundance variations for a reduced reaction rate of ^{26}Mg$(\alpha,n)^{29}$Si at $T_9 = 1.7$ after 1.4 days of burning time, analogous to figure 6.2.

6.1 ^{25}Mg(α,n)^{28}Si & ^{26}Mg(α,n)^{29}Si

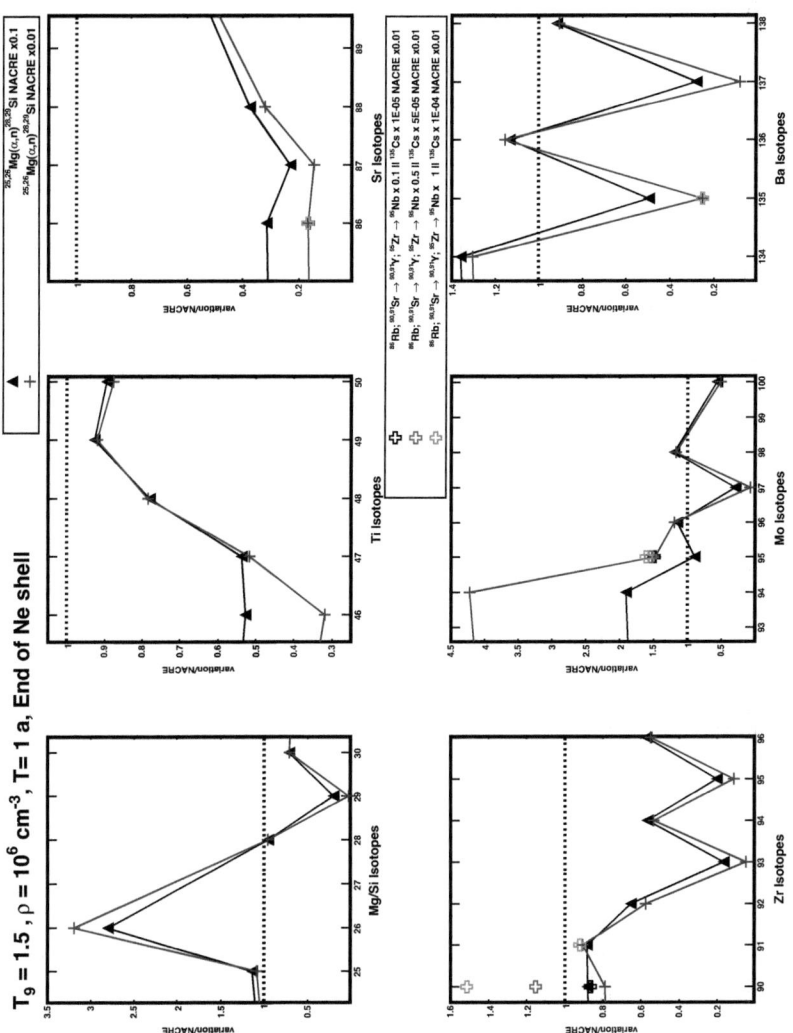

Figure 6.6: Illustrated are the isotopic abundance variations for the reduced reaction rates of ^{25}Mg(α,n)^{28}Si and ^{26}Mg(α,n)^{29}Si at $T_9 = 1.5$, analogous to figure 6.2.

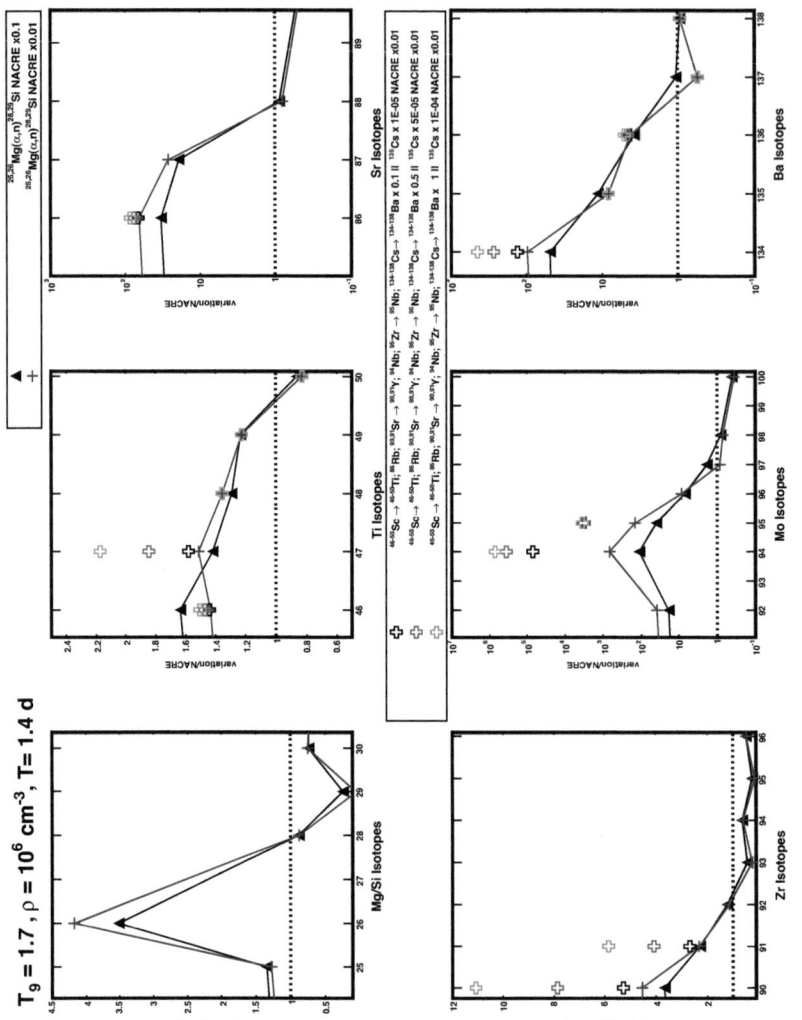

Figure 6.7: Illustrated are the isotopic abundance variations for the reduced reaction rates of $^{25}\mathrm{Mg}(\alpha,n)^{28}\mathrm{Si}$ and $^{26}\mathrm{Mg}(\alpha,n)^{29}\mathrm{Si}$ at $T_9 = 1.7$ after 1.4 days of burning time, analogous to figure 6.2.

The results from the calculations and the measured Ba signatures are compared in figures E.9 and E.10. The calculations reproduce the abundance of ^{138}Ba only at a temperature of $T_9 = 1.7$.

To explain the isotopic anomalies in grain KJB2-11-17-1, Hoppe et al. proposed an increase of the NACRE reaction rate of ^{26}Mg$(\alpha,n)^{29}$Si by a factor of two[54, 57]. As the experimental results for the reaction rate do show a decrease rather than an increase, a brief comparison with the calculated abundances is listed in table 6.3. Indeed, for the decreased reaction rates of ^{25}Mg$(\alpha,n)^{28}$Si and ^{26}Mg$(\alpha,n)^{29}$Si, the calculated ^{29}Si abundance seem to be in the same order for NACRE x 0.1 and $T_9 = 1.7$. But, in contrast, the ^{30}Si abundances differ substantially as compared to the ones found in grain KJB2-11-17-1. Calculations, in which only the reaction rate of ^{26}Mg$(\alpha,n)^{29}$Si was varied, were performed as well, but showed no improvement.

These results imply that the Si isotopes in this grain have not been synthesized during Ne burning alone. This conclusion is in accordance Hoppe et al. who note, that mixing between oxygen and neon zone has to be taken into account. However, the abundances calculated with the current data set only for the O/Ne zone also provide no match between the data sets.

On the other hand, due to the lower reaction rates, more ^{25}Mg and ^{26}Mg are left in the ashes of Ne burning, which then can be processed during further burning stages, such as oxygen burning or explosive burning phases. This could also have interesting implications for explosive nucleosynthesis calculations in massive stars and influence the final outcome of the resulting abundances. A more conclusive picture may be obtained by performing the same type of mixing calculations as Hoppe et al. did, but with the decreased reaction rates of ^{25}Mg$(\alpha,n)^{28}$Si and ^{26}Mg$(\alpha,n)^{29}$Si. This is beyond the scope of this thesis.

	$\delta\,^{29}$Si (‰)		$\delta\,^{30}$Si (‰)	
Hoppe et al. (1)	4929		2830	
Hoppe et al. (2)	10857		2830	
Hoppe et al. (3)	49		-168	
Hoppe et al. (4)	630		-168	
Ne burning calculations				
T_9	1.5	1.7	1.5	1.7
NACRE x 1	4538	5388	1549	3166
Variation of ^{26}Mg(α,n)^{29}Si only				
NACRE x 2.5	6153	6646	2343	4080
NACRE x0.1	344	847	734	1900
NACRE x0.01	-516	-353	674	1808
Variation of both reactions				
NACRE x0.1	193	711	936	2512
NACRE x0.01	-796	-700	929	2584

Table 6.3: Comparison of δ^iSi from this work and Hoppe et al.. (1) are the results only considering the O/Ne zone, (2) are the results only considering the O/Ne zone and increasing the ^{26}Mg(α,n)^{29}Si reaction rate by the suggested factor of 2.5, (3) are the results considering the mixing of all zones, (4) are the results considering the mixing of all zones and increasing ^{26}Mg(α,n)^{29}Si reaction rate by the factor 2.5. The results from (4) match the data of grain KJB2-11-17-1[54]. Below are the results for pure Ne shell burning with both the NACRE and lower reaction rates as indicated.

6.2 ^{18}O$(\alpha,n)^{21}$Ne

For the case of ^{18}O$(\alpha,n)^{21}$Ne, the reaction rate was adjusted and the latest results for ^{18}O$(\alpha,\gamma)^{22}$Ne by Dababneh et al. has been implemented as well[58].

In the He intershell region of AGB stars, ^{18}O is produced and rapidly destroyed during each thermal pulse (TP). At every TP the full He intershell becomes convective, mixing the ashes of hydrogen burning (which are rich in ^{14}N) into He burning environments (see section 2.2 and [9]). This leads to the reaction chain ^{14}N$(\alpha,\gamma)^{18}$F$(\beta^+)^{18}$O$(\alpha,\gamma)^{22}$Ne. During the short time scale of a TP (few hundred years), the ^{18}O abundance is defined by the ^{18}O$(\alpha,\gamma)^{22}$Ne reaction rate. At typical He burning temperatures ($T_9 = 0.25$), the results obtained here confirm the work by previous authors : the ^{18}O$(\alpha,n)^{21}$Ne reaction has a marginal influence on nucleosynthesis as compared to ^{18}O$(\alpha,\gamma)^{22}$Ne (see figure 5.38). At $T_9 = 0.25$, the reaction rate of ^{18}O$(\alpha,\gamma)^{22}$Ne is about three orders of magnitude higher than the reaction rate of ^{18}O$(\alpha,n)^{21}$Ne. Furthermore, this implies that even large variations of the ^{18}O$(\alpha,n)^{21}$Ne reaction rate would only lead to marginal changes in the nucleosynthesis since most of the ^{18}O is converted into ^{22}Ne.

Heck et al. observed ^{21}Ne excesses in presolar mainstream SiC grains from AGB stars that are due to cosmogenic processes and not to nucleosynthesis during the AGB phase[60]. To identify those excesses, the authors used as a baseline composition their prediction for the ^{21}Ne isotopic composition in the He shell of a M = 2 M$_\odot$ AGB star of solar metallicity. These values are compared in table 6.4 with the composition predicted from this study for a M = 1.5 M$_\odot$, half-solar metallicity AGB star. Both calculations show consistent results.

	^{20}Ne/^{22}Ne	^{21}Ne/^{22}Ne
Heck et al.	$6.5 \cdot 10^{-2}$	$5.9 \cdot 10^{-4}$
This work	$6.8 \cdot 10^{-2}$	$3.6 \cdot 10^{-4}$

Table 6.4: Listed are the results obtained in comparison to the results of Heck et al. [60].

A comparison of the calculated ^{21}Ne/^{22}Ne ratios compared to those actually measured in presolar grains (see figure 3.8), confirms that a strong additional cosmogenic contribution to the ^{21}Ne abundance is required.

The reaction rate obtained for ^{18}O$(\alpha,n)^{21}$Ne is up to a factor of 4 higher than the rate recommended by NACRE (see section 5.9). Implementing the new reaction rate and performing single-zone nucleosynthesis tests under He burning conditions showed only a marginal change in the ^{18}O abundance in the He shell, not affecting the conclusions obtained by Heck et al..

The reaction rates of ^{18}O$(\alpha,\gamma)^{22}$Ne and ^{18}F$(\alpha,p)^{21}$Ne have been recently determined to differ from previously recommended reaction rates[58, 92, 115]. In particular, using the upper limit of ^{18}F$(\alpha,p)^{21}$Ne, Karakas et al. calculated 21Ne/22Ne ratios that are about an order of magnitude higher than those given in Table 6.4. This may be less relevant for the grains analyzed by Heck et al., which contained larger contributions from cosmogenic Ne. However, as shown in figure 3.8, this could have an effect for the grain data obtained by Lewis et al., which were obtained on more typical SiC grains with a lower ratio of cosmogenic to AGB He shell Ne[59]. An intensive study regarding the new reaction rates and their effect on the synthesis of the neon isotopes is not within the scope of this thesis, but planned for the future.

The competition between the reactions ^{18}O$(\alpha,n)^{21}$Ne and ^{18}O$(\alpha,\gamma)^{22}$Ne does not play a role in more advanced burning phases (and therefore higher temperatures) as ^{18}O is completely destroyed in the early stages of He-burning.

Chapter 7

Conclusions

During the course of this thesis, a more complete picture on the impact of the reactions ^{25}Mg(α,n)^{28}Si, ^{26}Mg(α,n)^{29}Si and ^{18}O(α,n)^{21}Ne on stellar nucleosynthesis has been achieved.

^{25}Mg(α,n)^{28}Si & ^{26}Mg(α,n)^{29}Si For the reactions ^{25}Mg(α,n)^{28}Si and ^{26}Mg(α,n)^{29}Si it was shown that previous results were not accurate enough and needed further improvement. The improvements undertaken were the development of a more sophisticated neutron detection system and an advanced analysis method for the experimental data. Furthermore, various ways for the production of solid state targets were investigated and optimized.

Unfortunately, the measurements did not allow a successful separation of the experimental data from occuring background reactions at energies of astrophysical interest. It was shown that these background reactions result from impurities in the targets, not only resulting from carbon and oxygen impurites, but also from boron contaminations.

For future experiments on ^{25}Mg(α,n)^{28}Si and ^{26}Mg(α,n)^{29}Si, it will be crucial to remove these impurities during the target production process. A removal of the carbon and oxygen impurities is possible, but technically challenging. For example, one could operate the experimental beam line at vacuum levels of about 10^{-9} Pa to avoid an oxidation of the Mg target layer[99]. Sophisticated tests have to be undertaken for the removal of carbon impurities. Tode et al., for example, showed, that with an Ar excimer lamp it is possible to remove carbon contaminations on silicon wafers[117]. Along with the removal of boron, these measures would have to be performed in a vacuum system, which would allow the transfer of the target to the experimental beamline without leaving the vacuum.

An additional improvement would be the use of a germanium detector, operated in coincidence with a neutron detector. The germanium detector would constrain the experimental data by separating the γ-rays of interest from the ones released by background reactions. A disadvantage of the method would be the decreased efficiency and possible damage of the Ge crystal by the neutron irradiation. Additionally, the moderation times of the released neutrons have to be determined carefully to allow the set up of sufficient coincidence windows.

Concerning stellar nucleosynthesis, for the first time a quantitative assessment of the impact of the reaction rates has been achieved based on the obtained experimental results. The influence of the reaction ^{25}Mg(α,n)^{28}Si can be described as similar compared to the reaction ^{26}Mg(α,n)^{29}Si. While network calculations covered the Ne burning phase, the resulting seed composition for the following phases of stellar evolution is affected as well, which in turn influences also the isotopic abundance distribution in presolar SiC X grains. This aspect, also concerning the synthesis of the Si isotopes, has been addressed briefly but requires a more advanced study of stellar nucleosynthesis after Ne burning. Probably only the implementation of explosive synthesis will allow a final statement concerning the composition of presolar SiC X grains. Due to the lack of experimental information on the (α,γ) channel, competition effects between the (α,n) and (α,γ) remain unresolved.

^{18}O(α,n)^{21}Ne The experimental investigation of the ^{18}O(α,n)^{21}Ne reaction only showed a marginal influence of background reactions. For the first time, the experimental measurements of Denker were independently confirmed in their trend and were scaled by a factor of four[8]. A reduction of the carbon contamination would be desirable to resolve clearly the resonances and their parameters below $E_\alpha = 1100$ keV.

The impact of the reaction ^{18}O(α,n)^{21}Ne has been investigated with the newly obtained reaction rates as well. The competition with the ^{18}O(α,γ)^{22}Ne channel turned out to be critical at a lower stellar temperature than previously reported. It could be shown however, that at temperatures, at which the ^{18}O(α,n)^{21}Ne reaction starts to dominate over the ^{18}O(α,γ)^{22}Ne reaction, no ^{18}O would be available due to its exhaustion in the He-intershell phase. Resulting Ne isotopic abundances are similar to previously obtained values. This confirms that the observed ^{21}Ne excesses in large presolar SiC grains can be assigned to cosmogenic processes and do not have a stellar origin[60].

The challenges and results presented through the course of this thesis have revealed new insights concerning the investigation of neutron releasing reactions. Especially the development of a detection system and targets revealed new effects and sources of experimental uncertainties. To resolve these uncertainties, technologies from other fields, such as solid state physics, need to be implemented. Despite these challenges, it could be shown, how the investigated reactions influence the synthesis of elements in specific stellar enviroments and how sensitive the results are to nuclear physics data. The goal of assessing the impact of the investigated reactions on stellar nucleosynthesis was achieved by connecting the research performed experimentally with theoretical calculations and measurements on presolar grains. In the future, this approach will allow the quantitative description of the importance of nuclear reactions for stellar nucleosynthesis.

Appendix A

Beam Tuning Procedures

A.1 Energy Change

1. Calculate new NMR setting via energy calibration function :

$$\text{NMR} = 53.54 \cdot 2 \cdot \sqrt{E(\text{keV})}$$

2. Use coarse and fine of the master reference to reduce the magnetic field in the analyzing magnet.

   - As you go down in particle energy, you will need to keep the accelerator stabilizer balanced by either extending the corona points or reducing the terminal charge.

   - Roughly every 100 keV, adjust the focus of the ion source to maximize the beam current on the beamstop.

   - If you go below the desired target frequency, do not go up with the master reference. This will cause hysteresis effects. You must recycle the analyzing magnet.

3. At the desired NMR frequency :

   - Adjust the focus of the ion source to maximize beam current on the beam stop.

   - Reduce the setting for the Y-steerers (Y1 and Y2) simultaneously to maximize the current on the beamstop.

A.2 Tuning Process for α-Beam

- Balance the current reading on the defining slits by adjusting the X-steerers (X1 and X2).

- Adjust the focus of the ion sources to maximize the beam current on the beamstop.

- Adjust the Y-steerers such that Y-upper and Y-lower are reading a balanced current and that at the same time the entrance slits to the switching magnet read a balanced current.

- Change to the beam profile monitor located in the target room (see A.3).
- Turn down the quadrupoles in the target room.
- Use the beamline steerers X and Y (rastering offsets) and wobblers to get a maximum of 0.5 μA on each of the target slits (C,D,E and F on the aux. meter panel).
- Do not use beamline steerers X1, Y1 and the vertical steerer of the switching magnet.

A.3 Switch to Beam Profile Monitor (BPM) in Target Room

- Do not change channel positions while the BPMs are running.
- BPM setting for channel 2 : position 1 is the BMP in the accelerator vault while position 4 is the BPM located in the target room.
- Turn off BPM 2.
- Change channel 2 to desired position
- Turn on BPM 2.
- Adjust the gain of BPM2.

Appendix B

^{3}He Counter

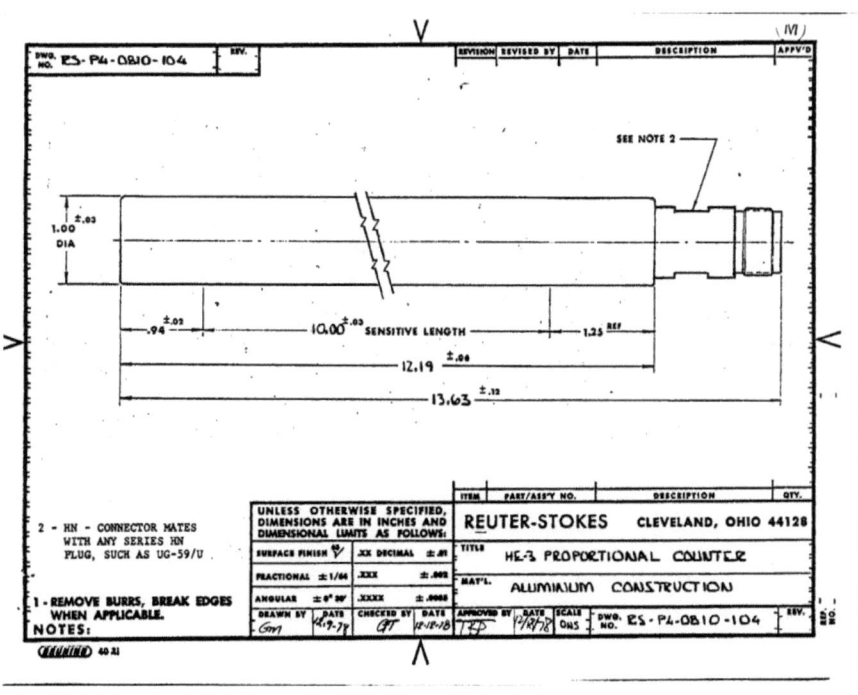

Figure B.1: Schematic illustration of a ^{3}He counter.

Appendix C

Correction Formalism

In this section, the method for the background correction described in section 5.4.2 is briefly discussed. The general formalism for the experimental yield etc. can be found in reference[27].
The experimental yield of a nuclear reaction is given by :

$$Y = \sigma \cdot \frac{\Delta}{\epsilon} \qquad (C.1)$$

where σ is the cross section, Δ the target thickness and ϵ the stopping power. As one analyzes the experimental data one has to determine whether a *thin* or a *thick* target is present. In general, a thick target can be assumed if the target thickness is at least 6 times the natural resonance width.
For a thin target the thin-target yield becomes :

$$Y(E_0) = \int_{E_0-\Delta}^{E_0} \frac{\sigma(E)}{\epsilon(E)} dE \qquad (C.2)$$

If $\Delta \gg \Gamma$, the yield becomes

$$Y(E_0) = \frac{\lambda^2}{2\pi}\omega\gamma \frac{M+m}{M}\frac{1}{\epsilon}\{arctan(\frac{E_0-E_R}{\Gamma/2}) - arctan(\frac{E_0-E_R-\Delta}{\Gamma/2})\} \qquad (C.3)$$

with M and m as the respective target and projectile masses. If a background component in form of a resonance peak occurs within the experimental data, one has to determine the thickness of the component first. By simply fitting the resonance peak with equation C.3 one can either determine the thickness, or if not possible, use the relation :

$$n_t = \frac{2}{(\omega\gamma)} \cdot \frac{1}{\lambda^2} \cdot I \qquad (C.4)$$

where n_t denotes the number of target atoms and I the integral over the peak region[118]. The measured experimental data at first is recorded in detected neutrons per incoming alpha particles. Since the background contributions already appear in this stage of the analysis, it was decided to apply the background corrections at the earliest stage. Generally, the corrected yield would then be :

$$Y_c = Y_{measured} - Y_{background} \qquad (C.5)$$

Since literature does not report experimental yields but cross sections (or S-factors), the literature data had to be convoluted into experimental yield data. This was done by calculating the experimental yield via the previous equations. For example, the stopping power was then varied assuming different compositions of the targets. The absolute cross section is defined as :

$$\sigma = \frac{N_n \epsilon}{N_\alpha \epsilon_{det} \Delta} \qquad (C.6)$$

with N_n being the detected neutrons, N_α the number of incoming α-particles and the detection efficiency ϵ_{det}. For the non-resonant part this allows then to caculate the experimental yield data for the background components. For the resonances occuring one has then to separate the non-resonant and resonant part and use the step function :

$$Y_{max}(\infty) = \frac{\lambda^2}{2} \omega\gamma \frac{M+m}{M} \frac{1}{\epsilon} \qquad (C.7)$$

as well, to determine the shape of the resonance peak for a given target thickness and resonance strength. One should take the natural enrichment of the background component into account as well.

Since the energies in the literature may deviate from the measured energies, an interpolation routine was implemented, which interpolates the background yield at the experimentally measured energies. The last step is then to apply an energy shift according to the peak position of the background resonance peak.

Appendix D

Analytic Expressions for the Reaction Rates

For each reaction rate an analytic expression was derived to be able to calculate the the reaction rate at any given temperature during the network calculations.
For the reactions ^{25}Mg(α,n)^{28}Si and ^{26}Mg(α,n)^{29}Si the formula derived by Caughlan and Fowler fitted the results best[48]. The results from the fits are summarized in tables D.1 and D.2.
The formula consists of the following terms for ^{25}Mg(α,n)^{28}Si :

$$N_A \langle \sigma v \rangle = p1 \cdot \frac{1}{gt9} \cdot (t9a)^{\frac{5}{6}} \cdot \frac{1}{T_9^{\frac{3}{2}}} \cdot e^{(p0 \cdot t9a^{\frac{1}{3}})} \quad \text{(D.1)}$$

with

$$t9a = T_9 \cdot \frac{1}{1 + p3 \cdot T_9} \quad \text{(D.2)}$$

$$gt9 = 1 + \frac{10}{3} \cdot e^{-\frac{p2}{T_9}} \quad \text{(D.3)}$$

and p_i being the fit parameters. The fit parameters have no physical meaning. For ^{26}Mg(α,n)^{29}Si the term gt9 becomes :

$$gt9 = 1 + 5 \cdot e^{-\frac{p2}{T_9}} \quad \text{(D.4)}$$

Parameter	$0.1 \leq T_9 \leq 1.1$	$1.1 \leq T_9 \leq 2.08$	$2.08 \leq T_9 \leq 10$
p0	59.76	32.75	45.73
p1	$5.68 \cdot 10^{18}$	$1.87 \cdot 10^{16}$	$1.83 \cdot 10^{15}$
p2	$6.71 \cdot 10^{14}$	-17.07	3.08
p3	-0.09	0.20	-0.02

Table D.1: Fit parameters for the reaction rate of ^{25}Mg(α,n)^{28}Si.

Parameter	$0.1 \leq T_9 \leq 1$	$1 \leq T_9 \leq 2.48$	$2.48 \leq T_9 \leq 10$
p0	52.46	93.84	20.79
p1	$5.03 \cdot 10^{17}$	$2.65 \cdot 10^{38}$	$3.57 \cdot 10^{9}$
p2	-2.67	7.42	-10.58
p3	0.03	0.39	-0.06

Table D.2: Fit parameters for the reaction rate of ^{26}Mg$(\alpha,n)^{29}$Si.

As the results for the ^{18}O$(\alpha,n)^{21}$Ne reaction bascially follow Denker's results, the same parametrization was chosen :

$$N_A \langle \sigma v \rangle = \frac{1}{1 + 9 \cdot 10^3 \cdot e^{-\frac{1}{2T_9}}} \cdot e^{-\frac{8.014}{T_9}} + \sum_{i=1}^{7} \frac{a_i}{T_9^{3/2}} \cdot e^{-\frac{b_i}{T_9}} \quad (D.5)$$

with a_i and b_i listed in tables D.3 and [8]. For the present data only the parameters a_i were varied within the fitting routines.

i	1	2	3	4	5	6	7
a_i	1.814	104.6	$3.225 \cdot 10^4$	$2.158 \cdot 10^6$	$7.8 \cdot 10^8$	$1.1 \cdot 10^{10}$	$5 \cdot 10^{10}$
b_i	8.2083	9.017	11.15	14.73	23.61	36.1	48.5

Table D.3: Fit parameters for the reaction rate of ^{18}O$(\alpha,n)^{21}$Ne from Denker[8].

i	1	2	3	4	5	6	7
a_i	$-4.60 \cdot 10^4$	$1.76 \cdot 10^5$	$-5.07 \cdot 10^5$	$1.12 \cdot 10^7$	$3.11 \cdot 10^9$	$2.92 \cdot 10^{11}$	$2.04 \cdot 10^{11}$
b_i	8.2083	9.017	11.15	14.73	23.61	36.1	48.5

Table D.4: Fit parameters for the reaction rate of ^{18}O$(\alpha,n)^{21}$Ne from the present work.

Appendix E

Nucleosynthesis Plots

E.1 Comparison to SiC X Data

The following plots show the achieved results from the network calculations in comparison to SiC X data. The abundance variations were calculated according to equation 6.1. Table E.1 shows the respective isotopes to which the abundances variations were normalized to.

Element	Normalization Isotope
Ti	^{48}Ti
Sr	^{86}Sr
Zr	^{94}Zr
Mo	^{96}Mo
Ba	^{135}Ba

Table E.1: List of for the elements considered in this discussion (Ti, Sr, Zr, Mo, Ba) and the respective isotopes to which the abundance variations are normalized to.

The effect by the radiogenic component on the abundance variation is shown as well, analogous to figures 6.2 - 6.7. For Ti, the results of the network calculations were in a range that did not allow a reasonable plotting together with the SiC X data (see figures E.1 and E.2).

Ti

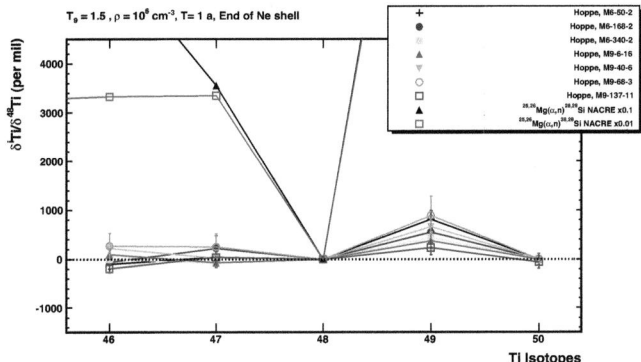

Figure E.1: Comparison of obtained Ti abundance variations (at $T_9=1.5$) to data from SiC X grains[107].

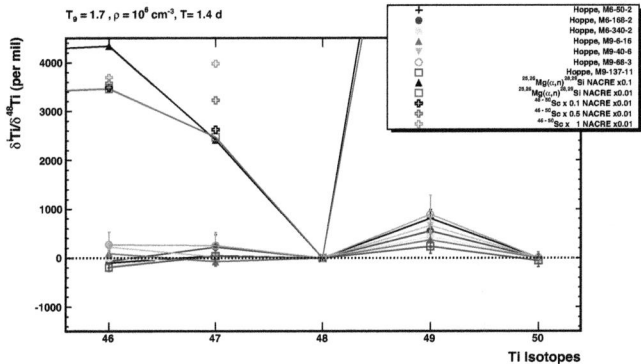

Figure E.2: Comparison of obtained Ti abundance variations (at $T_9=1.7$ after 1.4 days of burning time) to data from SiC X grains[107].

Sr

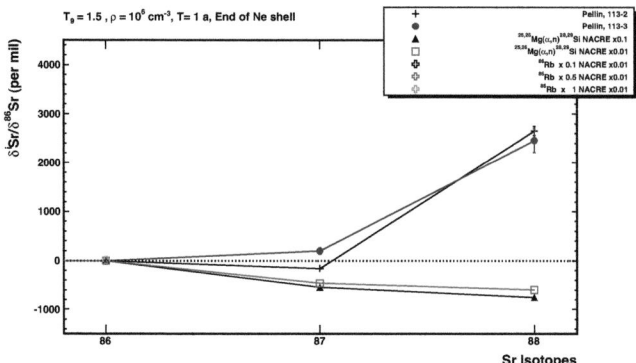

Figure E.3: Comparison of obtained Sr abundance variations (at $T_9=1.5$) to data from SiC X grains[112].

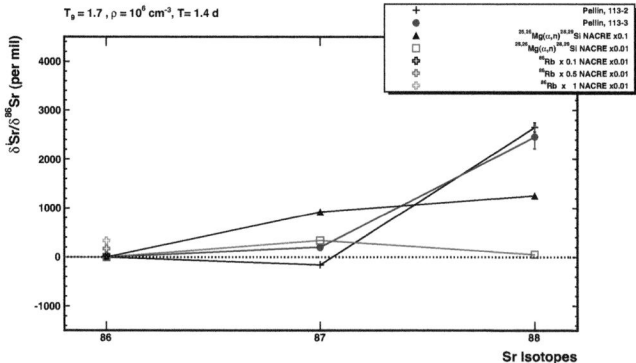

Figure E.4: Comparison of obtained Sr abundance variations (at $T_9=1.7$ after 1.4 days of burning time) to data from SiC X grains[112].

E.1 Comparison to SiC X Data

Zr

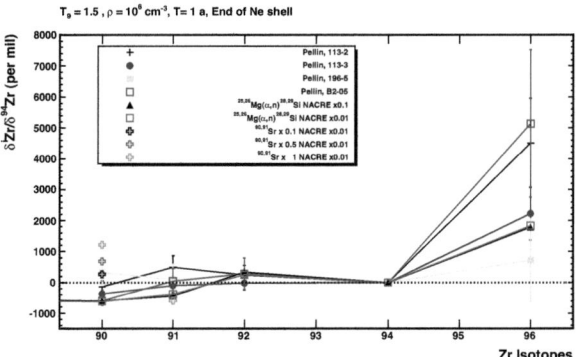

Figure E.5: Comparison of obtained Zr abundance variations (at $T_9=1.5$) to data from SiC X grains[112].

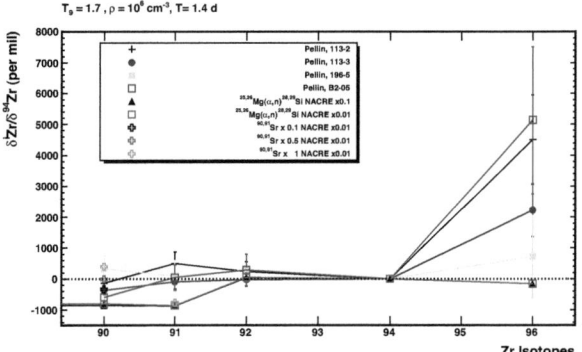

Figure E.6: Comparison of obtained Zr abundance variations (at $T_9=1.7$ after 1.4 days of burning time) to data from SiC X grains[112].

Mo

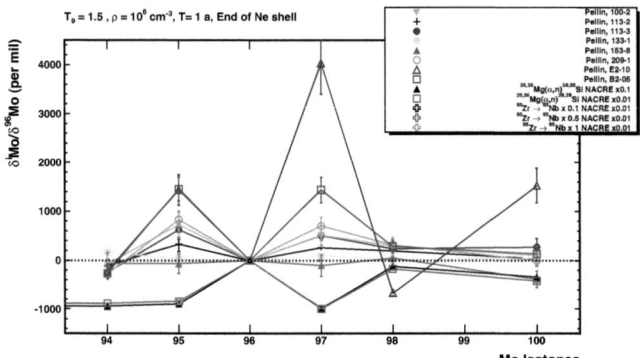

Figure E.7: Comparison of obtained Mo abundance variations (at T_9=1.5) to data from SiC X grains[112].

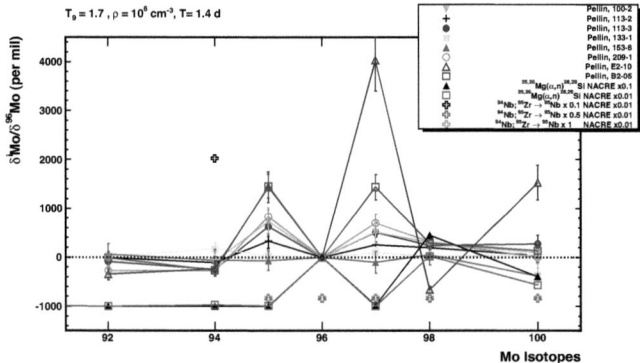

Figure E.8: Comparison of obtained Mo abundance variations (at T_9=1.7 after 1.4 days of burning time) to data from SiC X grains[112].

E.1 Comparison to SiC X Data

Ba

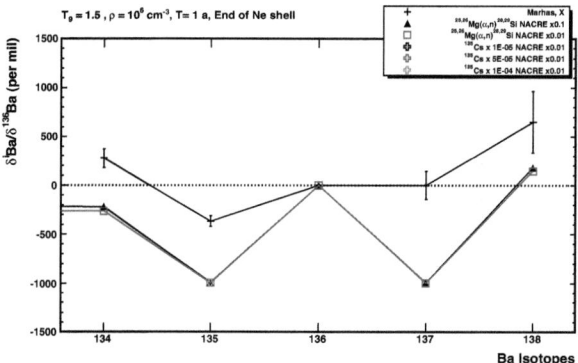

Figure E.9: Comparison of obtained Ba abundance variations (at T_9=1.5) to data from SiC X grains[113].

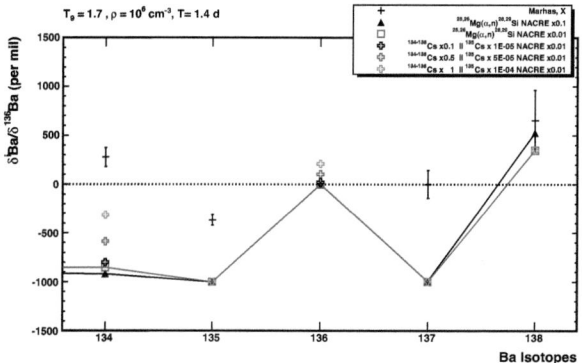

Figure E.10: Comparison of obtained Ba abundance variations (at T_9=1.7 after 1.4 days of burning time) to data from SiC X grains[113].

E.2 Abundance Evolution

The abundance of specific isotopes as a function of time is plotted in the following figures. For the isotopes ^{25}Mg and ^{26}Mg the abundance evolution is plotted in each plot to allow a comparison to the reaction rates of ^{25}Mg$(\alpha,n)^{28}$Si and ^{26}Mg$(\alpha,n)^{29}$Si. Note, once an abundance reaches 10^{-13} the isotope is regarded as completely destroyed. The abundance variations for $T_9 = 1.7$ are shown for one year burning time, but are only interesting with respect to nucleosynthesis for a burning time of 1.4 days.

E.2 Abundance Evolution

$T_9 = 1.5$

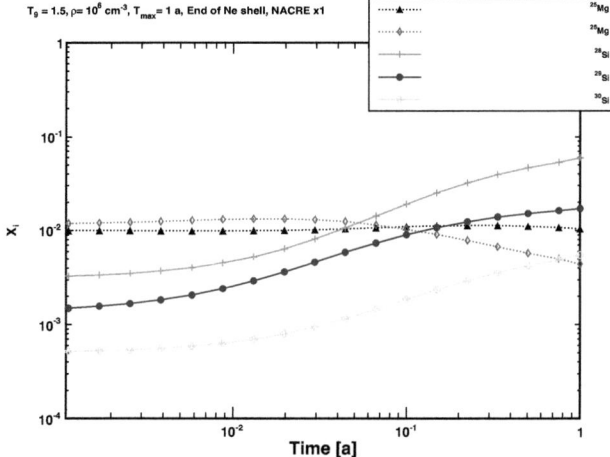

Figure E.11: Si abundance evolution at T_9=1.5 for NACREx1.

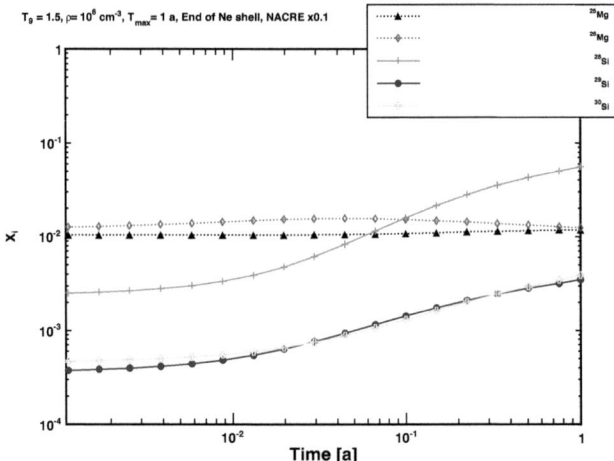

Figure E.12: Si abundance evolution at $T_9=1.5$ for NACREx0.1.

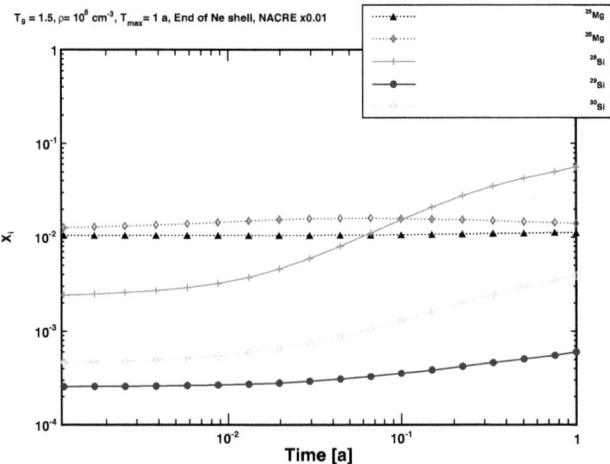

Figure E.13: Si abundance evolution at $T_9=1.5$ for NACREx0.01.

E.2 Abundance Evolution

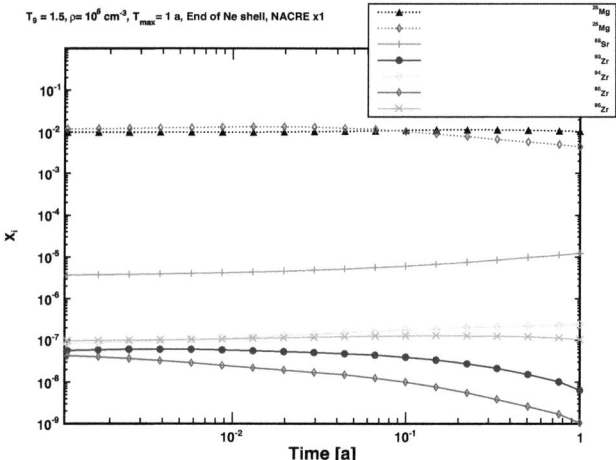

Figure E.14: Zr abundance evolution at $T_9=1.5$ for NACREx1.

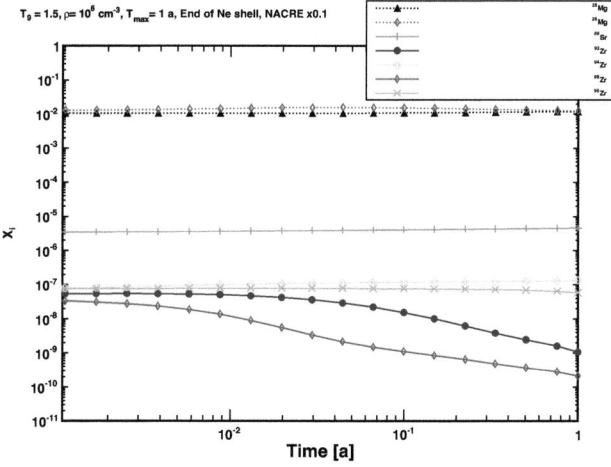

Figure E.15: Zr abundance evolution at $T_9=1.5$ for NACREx0.1.

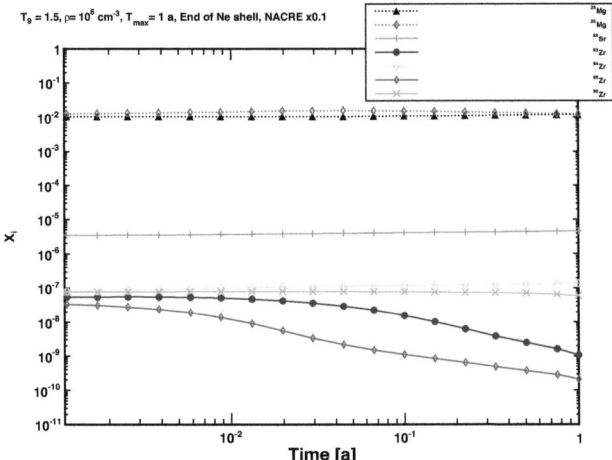

Figure E.16: Zr abundance evolution at $T_9=1.5$ for NACREx0.01.

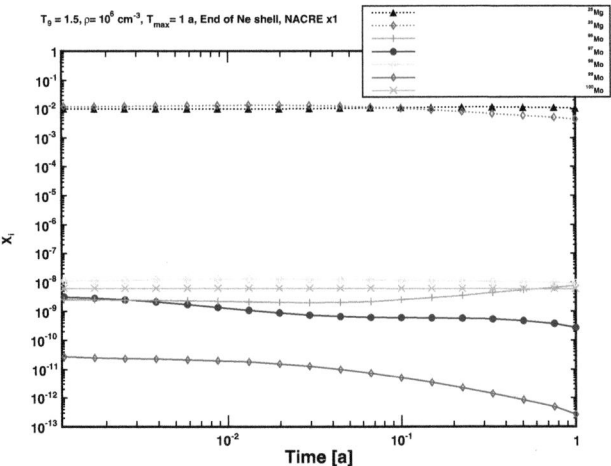

Figure E.17: Mo abundance evolution at $T_9=1.5$ for NACREx1.

E.2 Abundance Evolution

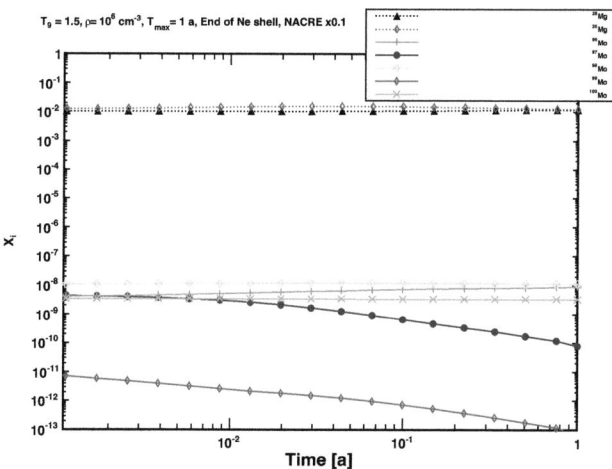

Figure E.18: Mo abundance evolution at T_9=1.5 for NACREx0.1.

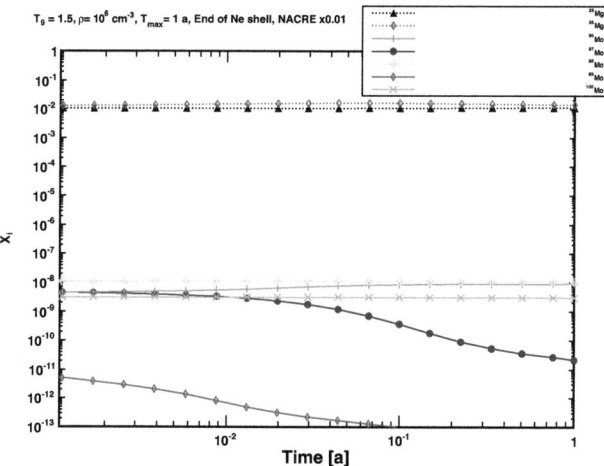

Figure E.19: Mo abundance evolution at T_9=1.5 for NACREx0.01.

$T_9 = 1.7$

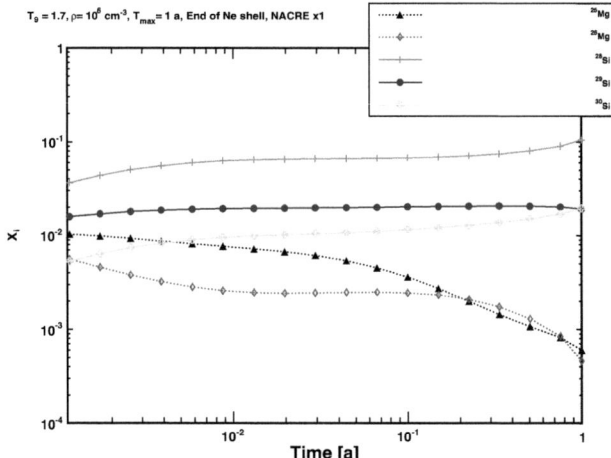

Figure E.20: Si abundance evolution at T_9=1.7 for NACREx1.

E.2 Abundance Evolution

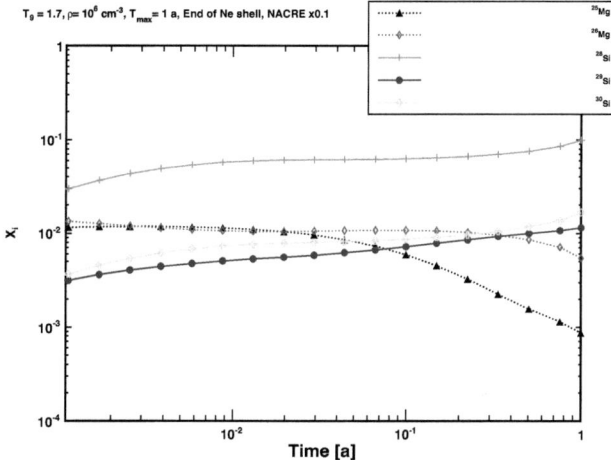

Figure E.21: Si abundance evolution at T_9=1.7 for NACREx0.1.

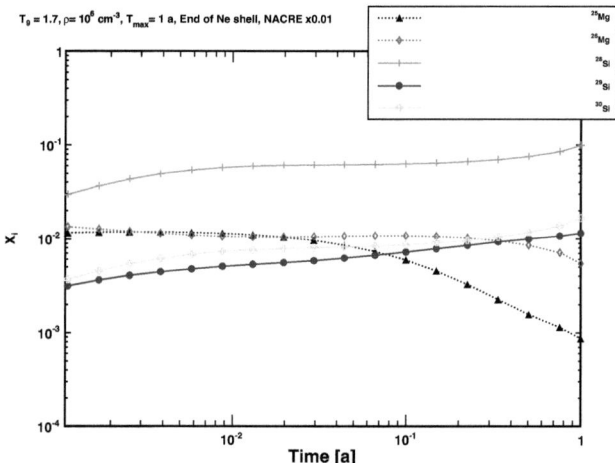

Figure E.22: Si abundance evolution at T_9=1.7 for NACREx0.01.

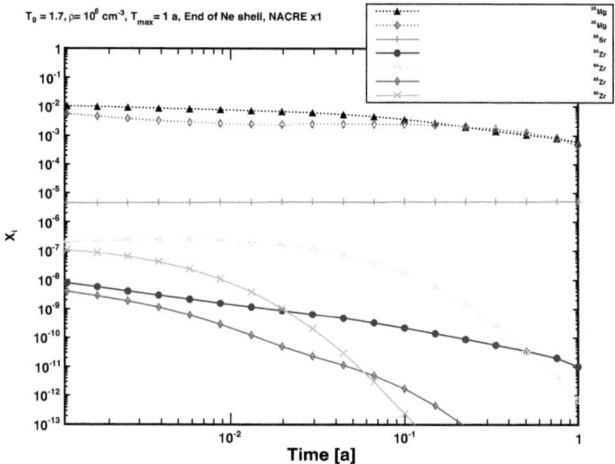

Figure E.23: Zr abundance evolution at $T_9=1.7$ for NACREx1.

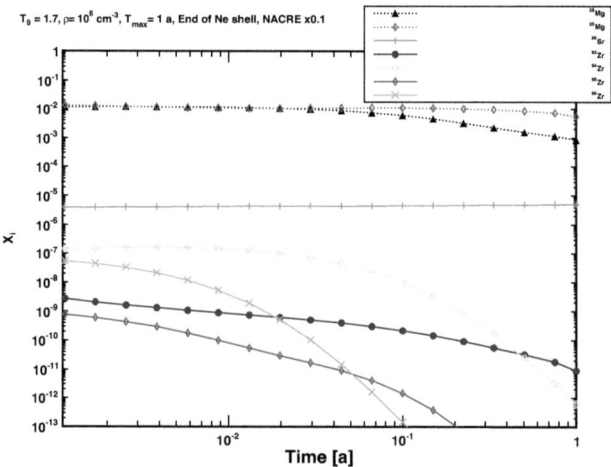

Figure E.24: Zr abundance evolution at $T_9=1.7$ for NACREx0.1.

E.2 Abundance Evolution

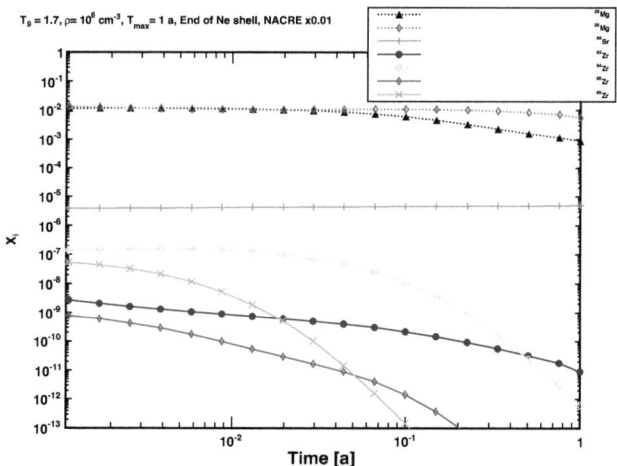

Figure E.25: Zr abundance evolution at T_9=1.7 for NACREx0.01.

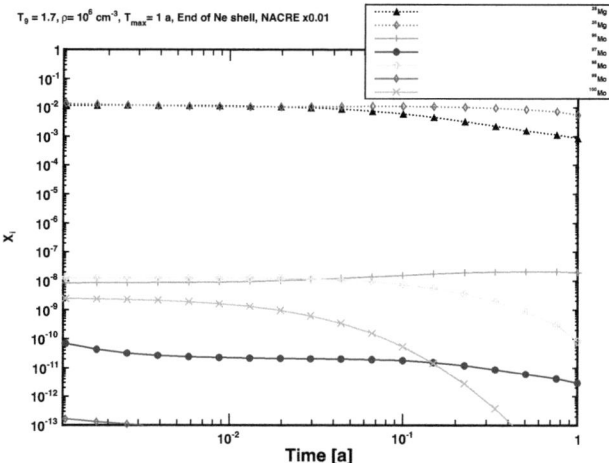

Figure E.26: Mo abundance evolution at T_9=1.7 for NACREx1.

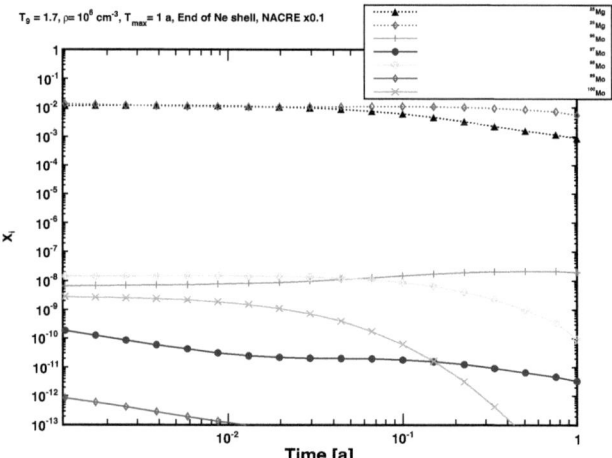

Figure E.27: Mo abundance evolution at $T_9=1.7$ for NACREx0.1.

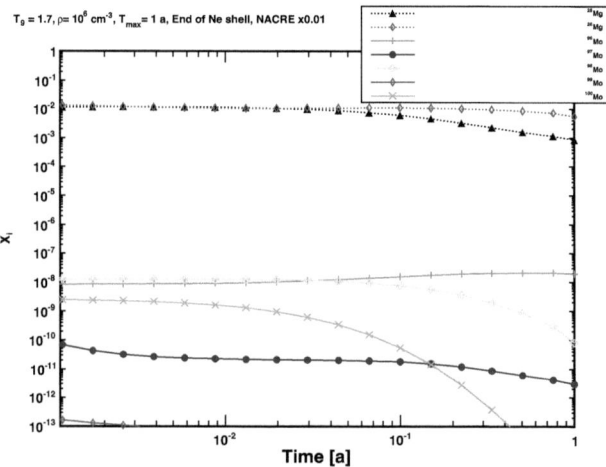

Figure E.28: Mo abundance evolution at $T_9=1.7$ for NACREx0.01.

Acknowledgements

First and foremost I want to thank Dr. Ulrich Ott for not only being officially being my advisor but also helping me with the struggles through the course of my Ph.D. studies. Prof. Dr. Karl-Ludwig Kratz and Prof. Dr. Klaus Wendt I want to thank to give me the unique opportunity to work on this Ph.D. project as well.

Prof. Dr. Michael Wiescher and Prof. Dr. Joachim Görres gave me the chance to work at the NSL at the University of Notre Dame. Through their support, I was given a deep insight into the field of experimental nuclear physics and was able to manage the challenges associated with this project.

Dr. Manoel Couder and Dr. Edward Stech have been a blessing. Without them I would be still stuck in GEANT, MCNP and the operators console of the NSL. Thousand thanks for sharing your experience and ideas with me.

Dr. Larry Lamm and his staff at the NSL I would like to thank for teaching me the ways of how things are kept in shape, machined, and how to maintain a healthy amount of humor..

Andreas Best deserves part of the credit as well, as he inherited the infamous neutron detector from me and supported me through the course of my work as well.

Countless sleepless nights, infinite amounts of calculations and stories I got to share with a passionate astrophysicist. Dr. Marco Pignatari! Thank you!

Shawn O'Brien and his wife Suzie I share my deepest respect with for welcoming me to their families and sharing their friendship with me.

P. J. LeBlanc, Ethan Uberseder, Beth Given and Brad Weldon I want to thank as well for supporting me during my time at Notre Dame and for becoming my friends.

Timo Griesel und Florian Schertz sei für die Unterstützung und Freundschaft schon seit Beginn unseres Studiums von ganzem Herzen gedankt.

Es gibt keine Worte, welche die Rolle meiner Eltern und meines Bruder während meiner Zeit als Doktorand beschreiben könnten. In tiefester Zuneigung : Danke!

Bibliography

[1] A. S. Eddington. The Internal Constitution of the Stars. *Science*, 52:233–240, 1920.

[2] E. M. Burbidge et al. Synthesis of the Elements in Stars. *Rev. Mod. Phys.*, 29:547–650, 1957.

[3] A. G. W. Cameron. Origin of Anomalous Abundances of the Elements in Giant Stars. *Astrophysical Journal*, 121:144, 1955.

[4] W. A. Fowler et al. Stellar Evolution and the Synthesis of the Elements. *Astrophysical Journal*, 122:271, 1955.

[5] R. Hirschi et al. NuGrid: S Process in Massive Stars. In *Nuclei in the Cosmos (NIC X)*, 2008.

[6] R. D. Hoffman et al. Nucleosynthesis below A = 100 in Massive Stars. *The Astrophysical Journal*, 549:1085–1092, 2001.

[7] M. Pignatari. *The Weak S-Process in Massive Stars*. PhD thesis, Universita Degli Studi Di Torino, 2006.

[8] A. Denker. *Drei Neutronenerzeugungsreaktionen in Sternen*. PhD thesis, University of Stuttgart, 1994.

[9] F. Herwig. Evolution of Asymptotic Giant Branch Stars. *Annual Review of Astronomy and Astrophysics*, 43:435–479, 2005.

[10] S. E. Woosley et al. The Evolution and Explosion of Massive Stars. *Rev. Mod. Phys.*, 74:1015–1071, 2002.

[11] M. Arnould and S. Goriely. The P-Process of Stellar Nucleosynthesis: Astrophysics and Nuclear Physics Status. *Physics Reports*, 384:1–84, 2003.

[12] C. Fröhlich et al. Neutrino-Induced Nucleosynthesis of A > 64 Nuclei: The νp Process. *Physical Review Letters*, 96:142502, 2006.

[13] D. D. Clayton. *Principles of Stellar Evolution and Nucleosynthesis*. The University of Chicago Press, 1968.

[14] D. D. Clayton et al. Neutron Capture Chains in Heavy Element Synthesis. *Annals of Physics*, 12:331 – 408, 1961.

[15] P. A. Seeger et al. Nucleosynthesis of Heavy Elements by Neutron Capture. *Astrophysical Journal Supplement Series*, 11:121, 1965.

[16] R. A. Ward and M. J. Newman. S-Process Studies - The Effects of a Pulsed Neutron Flux. *The Astrophysical Journal*, 219:195–212, 1978.

[17] F. Käppeler et al. S-Process Nucleosynthesis - Classical Approach and Asymptotic Giant Branch Models for Low-mass Stars. *The Astrophysical Journal*, 354:630–643, 1990.

[18] C. Arlandini et al. Neutron Capture in Low-Mass Asymptotic Giant Branch Stars: Cross Sections and Abundance Signatures. *The Astrophysical Journal*, 525:886–900, 1999.

[19] D. D. Clayton and M. E. Rassbach. Termination of the S-Process. *The Astrophysical Journal*, 148:69, 1967.

[20] F. Käppeler. The Origin of the Heavy Elements: The S-Process. *Progress in Particle and Nuclear Physics*, 43:419–483, 1999.

[21] K. Wisshak et al. Stellar Neutron Capture Cross Sections of the Nd Isotopes. *Phys. Rev. C*, 57:391–408, 1998.

[22] R. Gallino et al. Evolution and Nucleosynthesis in Low-Mass Asymptotic Giant Branch Stars. II. Neutron Capture and the S-Process. *The Astrophysical Journal*, 497:388, 1998.

[23] M. Pignatari et al. The Weak s-Process in Massive Stars and its Dependence on the Neutron Capture Cross Sections. *The Astrophysical Journal*, 710:1557–1577, 2010.

[24] O. Straniero et al. Evolution and Nucleosynthesis in Low-Mass Asymptotic Giant Branch Stars. I. Formation of Population I Carbon Stars. *The Astrophysical Journal*, 478:332, 1997.

[25] C. M. Raiteri et al. The Weak S-Component and Nucleosynthesis in Massive Stars. *The Astrophysical Journal*, 419:207, 1993.

[26] M. Pignatari. Private communications. 2007.

[27] C. E. Rolfs and William S. Rodney. *Cauldrons in the Cosmos*. The University of Chicago Press, 1988.

[28] G. Gamow. Zur Quantentheorie des Atomkernes. *Zeitschrift für Physik*, 51:204–212, 1928.

[29] E. E. Salpeter. Electrons Screening and Thermonuclear Reactions. *Australian Journal of Physics*, 7:373, 1954.

[30] F. Herwig et al. Nucleosynthesis Simulations for a Wide Range of Nuclear Production Sites from NuGrid. In *Nuclei in the Cosmos (NIC X)*, 2008.

[31] M. Pignatari et al. Complete Nucleosynthesis Calculations for Low-mass Stars from NuGrid. In *Nuclei in the Cosmos (NIC X)*, 2008.

[32] H. E. Suess. Chemical Evidence Bearing on the Origin of the Solar System. *Annual Review Astronomy and Astrophysics*, 3:217, 1965.

[33] A. G. W. Cameron. The Formation of the Sun and Planets. *Icarus*, 1:13–69, 1962.

[34] U. Ott. Interstellar grains in meteorites. *Nature*, 364:25–33, 1993.

[35] E. K. Zinner. Presolar Grains. *Treatise on Geochemistry*, 1:17–39, 2003.

[36] J. K. Bair and H. B. Willard. Level Structure in ^{22}Ne and ^{30}Si from the Reactions ^{18}O(α, n)^{21}Ne and ^{26}Mg(α,n)^{29}Si. *Phys. Rev.*, 128:299–304, 1962.

[37] J. P. Russell et al. Levels in ^{30}Si Excited by the ^{26}Mg(α,n)^{29}Si, ^{26}Mg(α,γ)^{30}Si and ^{26}Mg(α,α)^{26}Mg Reactions. *Nuclear Physics A*, 187:449 – 458, 1972.

[38] L. V. Namjoshi et al. Resonance Spectroscopy of the ^{30}Si Nucleus in the Excitation Energy Range 14.27 to 15.02 MeV. *Phys. Rev. C*, 13:915–921, 1976.

[39] O. Bassey Okon II. *Investigations in Radiative Alpha Capture in $^{29}Si,^{26}Mg,^{35}P,^{35}Cl$ and ^{37}Cl* . PhD thesis, State University of New York at Albany, 1973.

[40] L. Van der Zwan and K. W. Geiger. Cross Sections for the ^{25}Mg(α,n)^{28}Si Reaction at E$_\alpha$ < 4.8 MeV. *Nucl. Sci. Eng.*, 79:197–201, 1981.

[41] W. M. Howard et al. Nucleosynthesis of Rare Nuclei from Seed Nuclei in Explosive Carbon Burning. *The Astrophysical Journal*, 175:201, 1972.

[42] M. R. Anderson et al. ^{25}Mg(α,n)^{28}Si and ^{26}Mg(α,n)^{29}Si as Neutron Sources in Explosive Neon Burning. *Nuclear Physics A*, 405:170 – 178, 1983.

[43] S. Küchler. Untersuchung der Reaktionen ^{25}Mg(α,n)^{28}Si und ^{26}Mg(α,n)^{29}Si. Master's thesis, University of Stuttgart, 1990.

[44] J. K. Bair and F. X. Haas. Total Neutron Yield from the Reactions ^{13}C(α,n)^{16}O and ^{17,18}O(α, n)^{21}Ne. *Phys. Rev. C*, 7:1356–1364, 1973.

[45] O. Wieland. Untersuchung der Reaktionen ^{25}Mg(α,n)^{28}Si und ^{26}Mg(α,n)^{29}Si. Master's thesis, University of Stuttgart, 1995.

[46] R. Crowter. Master's thesis. Master's thesis, University of Surrey, 2007.

[47] S. E. Woosley et al. Semiempirical Thermonuclear Reaction-Rate Data for Intermediate-Mass Nuclei. *Atomic Data and Nuclear Data Tables*, 22:371 – 441, 1978.

[48] G. R. Caughlan and W. A. Fowler. Thermonuclear Reaction Rates V. *Atomic Data and Nuclear Data Tables*, 40:283, 1988.

[49] L. E. Brown and D. D. Clayton. SiC Particles from Asymptotic Giant Branch Stars - Mg Burning and the S-Process. *Astrophysical Journal Letters*, 392:L79–L82, 1992.

[50] P. Hoppe et al. Carbon, Nitrogen, Magnesium, Silicon and Titanium Isotopic Compositions of Single Interstellar Silicon Carbide Grains from the Murchison Carbonaceous Chondrite. *The Astrophysical Journal*, 430:870–890, 1994.

[51] E. Zinner et al. Silicon and Carbon Isotopic Ratios in AGB Stars: SiC Grain Data, Models, and the Galactic Evolution of the Si Isotopes. *The Astrophysical Journal*, 650:350–373, 2006.

[52] S. Mostefaoui and P. Hoppe. Discovery of Abundant In Situ Silicate and Spinel Grains from Red Giant Stars in a Primitive Meteorite. *The Astrophysical Journal Letters*, 613:L149–L152, 2004.

[53] M. Lugaro et al. Si Isotopic Ratios in Mainstream Presolar SiC Grains Revisited. *The Astrophysical Journal*, 527:369–394, 1999.

[54] P. Hoppe et al. An Unusual Presolar Silicon Carbide Grain From a Supernova : Implications for the Production of ^{29}Si in Type II Supernovae. *The Astrophysical Journal Letters*, 691, 2009.

[55] T. Rauscher et al. Nucleosynthesis in Massive Stars with Improved Nuclear and Stellar Physics. *The Astrophysical Journal*, 576:L20–L23, 2002.

[56] W. A. Fowler et al. Thermonuclear Reaction Rates II. *Annual Review of Astronomy and Astrophysics*, 13:69, 1975.

[57] C. Angulo et al. A Compilation of Charged-Particle Induced Thermonuclear Reaction Rates. *Nuclear Physics A*, 656:3 – 183, 1999.

[58] S. Dababneh et al. Stellar He Burning of ^{18}O : A Measurement of Low-Energy Resonances and their Astrophysical Implications. *Phys. Rev. C*, 68:025801, 2003.

[59] R. S. Lewis et al. Interstellar Grains in Meteorites: II. SiC and its Noble Gases. *Geochimica et Cosmochimica Acta*, 58:471 – 494, 1994.

[60] P. R. Heck et al. Interstellar Residence Times of Presolar SiC Dust Grains from the Murchison Carbonaceous Meteorite. *The Astrophysical Journal*, 698:1155–1164, 2009.

[61] J. C. Overley et al. The Energy Calibration of Tandem Accelerators. *Nuclear Instruments and Methods*, 68:61 – 69, 1969.

[62] C. H. Johnson et al. Thresholds for (p,n) Reactions on 26 Intermediate-Weight Nuclei. *Phys. Rev.*, 136:B1719–B1729, 1964.

[63] National Nuclear Data Center. Evaluated Nuclear Data Files - ENDF and JENDL. *http://www.nndc.bnl.gov/exfor/endf00.jsp*.

[64] A. O. Hanson and J. L. McKibben. A Neutron Detector Having Uniform Sensitivity from 10 keV to 3 MeV. *Phys. Rev.*, 72:673–677, 1947.

[65] L.V. East and R.B. Walton. Polyethylene Moderated ^{3}He Neutron Detectors. *Nuclear Instruments and Methods*, 72:161 – 166, 1969.

[66] B. Holmqvist and E. Ramström. A High Efficiency 4π Neutron Detector. *Nuclear Instruments and Methods in Physics Research*, 188:153 – 157, 1981.

[67] L. B. Borst. Low-Temperature Neutron Moderation. *Phys. Rev. Lett.*, 4:131, 1960.

[68] K. Wolke. Helium burning of ^{22}Ne. *Zeitschrift für Physik A Hadrons and Nuclei*, 334:491–510, 1989.

[69] V. Harms et al. Properties of ^{22}Ne$(\alpha,n)^{25}$Mg resonances. *Phys. Rev. C*, 43:2849–2861, 1991.

[70] H. W. Drotleff. *Untersuchung der S-Prozess Neutronenquelle $^{22}Ne(\alpha, n)^{25}Mg$*. PhD thesis, Ruhr-Universität Bochum, 1992.

[71] M. Jäger. *Die Einfangreaktion $^{22}Ne(\alpha,n)^{25}Mg$ -die Hauptneutronenquelle in Massiven Sternen*. PhD thesis, Universität Stuttgart, 2001.

[72] S. Agostinelli et al. GEANT4 – a Simulation Toolkit. *Nuclear Instruments and Methods in Physics Research Section A: Accelerators, Spectrometers, Detectors and Associated Equipment*, 506:250–303, 2003.

[73] Los Alamos National Laboratory. MCNP - A General Monte Carlo N-Particle Transport Code - Version 5. *http://mcnp-green.lanl.gov/index.html*.

[74] GEANT4 - User Documentation and Forum. *http://geant4.web.cern.ch/geant4/support/*.

[75] T. Koi. GEANT4 - Thermal Neutron Scattering in GEANT4. *http://geant4hadronics.wikispaces.com/Thermal+Neutron+Scattering+in+Geant4*.

[76] The Root Team CERN. ROOT - An Object Oriented Data Analysis Framework. *http://root.cern.ch/*.

[77] J. L. Zyskind et al. Competition effects in proton-induced reactions on ^{51}V. *Nuclear Physics A*, 343:295 – 314, 1980.

[78] K. K. Harris et al. The ^{51}V(p,n)^{51}Cr reaction as a neutron source of known intensity. *Nuclear Instruments and Methods*, 33:257 – 260, 1965.

[79] M. Heil et al. The ^{13}C(α,n) Reaction and its Role as a Neutron Source for the s Process. *Physical Review C (Nuclear Physics)*, 78:025803, 2008.

[80] M. Heagney. Reduction Techniques for Isotopic Materials. International Nuclear Target Development Society, Los Alamos National Laboratory, 1976.

[81] S. Takayanagi et al. On the Preparation of Magnesium Targets from MgO. *Nuclear Instruments and Methods*, 45:345 – 346, 1966.

[82] S.H. Maxman. Target Preparation Techniques. *Nuclear Instruments and Methods*, 50(1):53 – 60, 1967.

[83] R. B. Weinberg et al. Deuteron Stripping Reaction on ^{25}Mg. *Phys. Rev.*, 133:B884–B892, 1964.

[84] A. H. F. Muggleton. Preparation of Thin Nuclear Targets. *Journal of Physics E: Scientific Instruments*, 12:780, 1979.

[85] G. M. Hinn et al. Production of Thick Elemental Low-Oxygen Content ^{nat,26}Mg from ^{nat,26}MgO. *Nuclear Instruments and Methods in Physics Research Section A: Accelerators, Spectrometers, Detectors and Associated Equipment*, 227:434 – 436, 1984.

[86] D.A Vermilyea. The formation of anodic oxide films on tantalum in non-aqueous solutions. *Acta Metallurgica*, 2:482 – 486, 1954.

[87] A. Best. unpublished results. 2009.

BIBLIOGRAPHY 163

[88] A. Best. *to be published*. PhD thesis, University of Notre Dame.

[89] E. Strandberg et al. ^{24}Mg$(\alpha,\gamma)^{28}$Si Resonance Parameters at Low Alpha-Particle Energies. *Physical Review C (Nuclear Physics)*, 77:055801, 2008.

[90] C. E. Aalseth et al. Ultra-Low-Background Copper Production and Clean Fabrication. volume 785, pages 170–176. AIP, 2005.

[91] E.W. Hoppe et al. Cleaning and Passivation of Copper Surfaces to Remove Surface Radioactivity and Prevent Oxide Formation. *Nuclear Instruments and Methods in Physics Research Section A: Accelerators, Spectrometers, Detectors and Associated Equipment*, 579:486 – 489, 2007. Proceedings of the 11th Symposium on Radiation Measurements and Applications.

[92] H.Y. Lee. *The $^{18}F(\alpha,p)^{21}Ne$ Reaction and Its Astrophysical Implications*. PhD thesis, University of Notre Dame du Lac, 2006.

[93] J.F. Ziegler et al. *The Stopping and Range of Ions in Solids*. Pergamon Press, 1985.

[94] X. Xu et al. Interfacial Reactions between Oxide Films and Refractory Metal Substrates. *Langmuir*, 12:4877?4881, 1996.

[95] Albert E. Evans Jr. Energy dependence of the response of a ^{3}He long counter. *Nuclear Instruments and Methods in Physics Research*, 199:643 – 644, 1982.

[96] K. K. Sekharan et al. A Neutron Detector for Measurement of Total Neutron Production Cross Sections. *Nuclear Instruments and Methods*, 133:253 – 257, 1976.

[97] P. Hosmer. *Beta-Decay Studies of ^{78}Ni and Other Neutron-Rich Nuclei in the Astrophysical R-Process*. PhD thesis, Michigan State University, 2005.

[98] S. Harissopulos et al. Cross Section of the ^{13}C$(\alpha,n)^{16}$O Reaction: A Background for the Measurement of Geo-Neutrinos. *Phys. Rev. C*, 72:062801, 2005.

[99] M. Kurth. *The Initial Oxidation Of Magnesium*. PhD thesis, Universität Stuttgart, 2004.

[100] E. Raub et al. Inkorporation kohlenstoffhaltiger Fremdstoffe bei der galvanischen Abscheidung von Gold- und Goldlegierungsüberzügen in sauren und alkalischen Cyanidbädern. *Materials and Corrosion/Werkstoffe und Korrosion*, 23:643–647, 1972.

[101] T. R. Wang et al. ^{11}B+α Reaction Rates and Primordial Nucleosynthesis. *Phys. Rev. C*, 43:883–896, Feb 1991.

[102] P. M. Endt. Energy Levels of A = 21-44 Nuclei (VII). *Nuclear Physics A*, 521, 1990.

[103] P. M. Endt. Supplement to Energy Levels of A = 21-44 Nuclei (VII). *Nuclear Physics A*, 633:1 – 220, 1998.

[104] R.B. Firestone. Nuclear Data Sheets for A = 22. *Nuclear Data Sheets*, 106:1 – 88, 2005.

[105] T. Rauscher et al. Astrophysical Reaction Rates From Statistical Model Calculations. *Atomic Data and Nuclear Data Tables*, 75:1 – 351, 2000.

[106] S. W. J. Colgan et al. Day 640 infrared line and continuum measurements: Dust formation in SN 1987A. *The Astrophysical Journal*, 427:874–888, 1994.

[107] P. Hoppe and A. Besmehn. Evidence for Extinct Vanadium-49 in Presolar Silicon Carbide Grains from Supernovae. *The Astrophysical Journal Letters*, 576:L69–L72, September 2002.

[108] S. Amari, P. Hoppe, E. Zinner, and R. S. Lewis. Trace-element concentrations in single circumstellar silicon carbide grains from the Murchison meteorite. *Meteoritics*, 30:679, 1995.

[109] K. Lodders and B. Fegley, Jr. The origin of circumstellar silicon carbide grains found in meteorites. *Meteoritics*, 30:661, November 1995.

[110] J. Spyromilio, R. A. Stathakis, and G. R. Meurer. Clumping and Smallscale Mixing in Supernova 1987A. *Monthly Notices of the Royal Astronomical Society*, 263:530, 1993.

[111] E. Anders and N. Grevesse. Abundances of the Elements: Meteoritic and Solar. *Geochimica et Cosmochimica Acta*, 53:197 – 214, 1989.

[112] M. J. Pellin et al. Heavy Metal Isotopic Anomalies in Supernovae Presolar Grains. In *37th Annual Lunar and Planetary Science Conference*, volume 37, page 2041, 2006.

[113] K. K. Marhas et al. NanoSIMS studies of Ba isotopic compositions in single presolar silicon carbide grains from AGB stars and supernovae. *Meteoritics and Planetary Science*, 42:1077–1101, 2007.

[114] B. S. Meyer et al. Molybdenum and Zirconium Isotopes from a Supernova Neutron Burst. *The Astrophysical Journal Letters*, 540:L49–L52, 2000.

[115] A. I. Karakas et al. The Impact of the ^{18}F$(\alpha,p)^{21}$Ne Reaction on Asymptotic Giant Branch Nucleosynthesis. *The Astrophysical Journal*, 676:1254–1261, 2008.

[116] M. Pignatari et al. in preparation. 2010.

[117] M. Tode et al. Removal of Carbon Contamination on Si Wafers with an Excimer Lamp. *Metallurgical and Materials Transactions A*, 38:596–598, 2007.

[118] D. C. Powell et al. Low-Energy Resonance Strengths for Proton Capture on Mg and Al Nuclei. *Nuclear Physics A*, 644:263 – 276, 1998.

Die VDM Verlagsservicegesellschaft sucht für wissenschaftliche Verlage abgeschlossene und herausragende

Dissertationen, Habilitationen, Diplomarbeiten, Master Theses, Magisterarbeiten usw.

für die kostenlose Publikation als Fachbuch.

Sie verfügen über eine Arbeit, die hohen inhaltlichen und formalen Ansprüchen genügt, und haben Interesse an einer honorarvergüteten Publikation?

Dann senden Sie bitte erste Informationen über sich und Ihre Arbeit per Email an *info@vdm-vsg.de*.

Sie erhalten kurzfristig unser Feedback!

VDM Verlagsservicegesellschaft mbH
Dudweiler Landstr. 99   Telefon +49 681 3720 174
D - 66123 Saarbrücken   Fax   +49 681 3720 1749
www.vdm-vsg.de

Die VDM Verlagsservicegesellschaft mbH vertritt

Printed by Books on Demand GmbH, Norderstedt / Germany